Applications of Big Data and Artificial Intelligence in Smart Energy Systems
Volume 2
Energy Planning, Operations, Control and Market Perspectives

Editors

Neelu Nagpal
Maharaja Agrasen Institute of Technology, India

Hassan Haes Alhelou
Tishreen University, Syria

Pierluigi Siano
University of Salerno, Italy

Sanjeevikumar Padmanaban
Aalborg University, Denmark

D. Lakshmi
VIT Bhopal University, India

River Publishers

Routledge
Taylor & Francis Group
NEW YORK AND LONDON

Published 2023 by River Publishers
River Publishers
Alsbjergvej 10, 9260 Gistrup, Denmark
www.riverpublishers.com

Distributed exclusively by Routledge
605 Third Avenue, New York, NY 10017, USA
4 Park Square, Milton Park, Abingdon, Oxon OX14 4RN

Applications of Big Data and Artificial Intelligence in Smart Energy Systems / by Neelu Nagpal, Hassan Haes Alhelou, Pierluigi Siano, Sanjeevikumar Padmanaban, D. Lakshmi.

Routledge is an imprint of the Taylor & Francis Group, an informa business

ISBN 978-87-7022-827-5 (hardback)
ISBN 978-87-7022-995-1 (paperback)
ISBN 978-10-0096-397-7 (online)
ISBN 978-10-0344-086-4 (master ebook)

While every effort is made to provide dependable information, the publisher, authors, and editors cannot be held responsible for any errors or omissions.

Contents

2 Applications of Artificial Intelligence in Intelligent Combustion and Energy Storage Technologies   **27**

Kannan Chakrapani, Thiyagarajan Kavitha, Mohamed Iqubal Safa, Muniyegowda Kempanna, and Bharathi Chakrapani

3 Sustainable Smart Energy Systems and Energy Preservation Strategies in Intelligent Transportation Sectors   **47**
Bharathi Chakrapani, Kannan Chakrapani, Thiyagarajan Kavitha, Mohamed Iqubal Safa, and Muniyegowda Kempanna

9 Artificial Neural Network and Forecasting Major Electricity Markets   **193**

Vaibhav Aggarwal, Sudhi Sharma, and Adesh Doifode

Preface

A smart energy system is a strategy for coordinating multiple sources of power generation, power utilization, and power storage technologies to uncover synergies in order to provide an optimal solution for each individual sector as well as the overall energy system. The development of smart energy systems necessitates technological advancements for better automation, communication, and information technology systems to process, monitor, control, and automate power flows from points of generation to the points of consumption. In this paradigm, the incorporation of Artificial Intelligence (AI) technologies and IoT in energy systems has the potential to reduce energy cost, waste, and accelerate the global implementation of clean renewable energy sources. AI can be used to benefit all aspects of power system planning, operation, and control in addition to managing decentralized networks during the transition to future renewable and sustainable energy systems.

This book discusses various aspects of energy systems related issues, which exist in the conventional electrical sector, introducing requirements, various components/structure, features, and configuration of future energy networks under smart scenarios. This book provides in-depth information on a variety of areas of energy systems, as well as infrastructure development and maintenance. The book's following section delves into the fundamentals of AI and big data as well as numerous approaches to applying these ideas to real-world challenges, and case studies on energy planning, monitoring, control, and automation in real time.

Implementing big data science for better and safer operation is possible in the context of future smart energy systems' digitization and automation. Smart plugs, switches, and devices, smart grids, smart appliances, phase measurement, field measurement, RTUs, sensors mounted on grid-level equipment (e.g., transformers and network switches), asset inventory, SCADA system, Geographic Information System (GIS), weather data, traffic data, and social media are all expected to become massive data sources. This book covers real-time monitoring, control, and automation utilizing AI to access and extract data features. It will therefore include the approach used to detect voltage instability, margin prediction, real-time fault threshold computation, nonstationary faults, line outage detection, and expedited control and planning for energy restoration and protection.

This book will examine contemporary breakthroughs that use advanced technologies such as IoT, blockchain, cloud computing, and others to build smart grids that are efficient, safe, and stable. Suitable AI technology development is required for appropriate applications of various domains of energy systems, and this chapter seeks to be receptive to smart energy solutions employing data sciences and appropriate AI models. Numerous case studies are presented in the energy sector, including but not limited to the use of intelligent techniques and subfields such as machine learning, big data, blockchain, IoT, cloud computing, computer vision, and neural networks in networking, manufacturing, management, transportation, and shipping to construct energy-efficient and sustainable structures. Furthermore, the detailed state of the art and case studies with numerous experimental results and associated simulations contribute significantly to the high quality of the book.

This book contains two volumes. Volume 1 of the book has covered the framework of smart energy systems with the incorporation of artificial intelligence techniques and cutting-edge technologies such as IoT, Big Data analytics, and blockchain for demand prediction, decision-making processes, policy, and energy management, real-time energy monitoring, control and automation, and implementation.

Volume 2 contains nine chapters that provide the applications of Artificial Intelligence (AI) and case studies that are associated with the energy industry. This covers the use of AI for storage and preservation technologies for intelligent transportation and utilities, mitigating power quality challenges, and forecasting energy market. The advanced technologies in smart energy systems covered in this book include electro mobility, virtual power plants, digital security solutions, distribution automation, and marketing solutions in the energy industry. Further, problems in depth, solution approaches, implementation, and realistic case studies are also discussed.

Editors:

Neelu Nagpal
Maharaja Agrasen Institute of Technology, India

Hassan Haes Alhelou
Tishreen University, Syria

Pierluigi Siano
University of Salerno, Italy

Sanjeevikumar Padmanaban
Aalborg University, Denmark

D. Lakshmi
VIT Bhopal University, India

List of Figures

List of Tables

List of Contributors

Aggarwal, Vaibhav, *O.P. Jindal Global University, India*

Arumugam, Dhanasekaran, *Center for Energy Research, Department of Electrical and Electronics Engineering, Chennai Institute of Technology, India*

Badi, Manjulata, *Department of Electrical and Electronics Engineering, Alliance University, India*

Bhatt, Urvi Y., *Department of Computer Science and Engineering, CHARUSAT-DEPSTAR, India*

Chakrapani, Bharathi, *School of Computer Science and Engineering (SCOPE), Vellore Institute of Technology, India*

Chakrapani, Kannan, *Department of Information Technology, School of Computing, SASTRA Deemed University, India*

Doifode, Adesh, *Institute for Future Education, Entrepreneurship and Leadership, India*

Iqubal Safa, Mohamed, *Department of IT, SoC, SRM Institute of Science and Technology, India*

Jegadeesan, Vishnupriyan, *Center for Energy Research, Department of Electrical and Electronics Engineering, Chennai Institute of Technology, India*

Kassarwani, Neelam, *Department of Electrical Engineering, MAIT, India*

Kavitha, Thiyagarajan, *Department of Computer Science & Engineering, Koneru Lakshmaiah Education Foundation, India*

Kempanna, Muniyegowda, *Department of Computer Science & Engineering, BIT, Visvesvaraya Technological University (VTU), India*

Lohia, Agrima, *Area-Information Systems, IIM-Ahmedabad, India*

Mahapatra, Sheila, *Department of Electrical and Electronics Engineering, Alliance University, India*

Mishra, Jahnvi Rajiv, *Center for Energy Research, Department of Electrical and Electronics Engineering, Chennai Institute of Technology, India*

Palanikumarasamy, Vijay, *Center for Energy Research, Department of Electrical and Electronics Engineering, Chennai Institute of Technology, India*

Paul, Ajay John, *School of Mechanical Engineering, Kyungpook National University, South Korea*

Raj, Saurav, *Department of Electrical Engineering, Institute of Chemical Technology, Marathwada Campus, India*

Raval, Aanal, *Area-Information Systems, IIM-Ahmedabad, India*

Sharma, Sudhi, *Fortune Institute of International Business, India*

Shekarappa, G. Swetha, *Department of Electrical and Electronics Engineering, Alliance University, India*

Singh, Alka, *Department of Electrical Engineering, DTU, India*

Stephen, Christopher, *Department of Mechanical Engineering, Vel Tech Rangarajan Dr. Sagunthala R & D Institute of Science and Technology, India; TARE Fellow, National Institute of Solar Energy, India through Science and Engineering Research Board (SERB), India*

List of Abbreviations

ACF	Auto-correlation function
ADALINE	Adaptive linear neuron or adaptive linear element
ADF	Augmented Dickey–Fuller
AEST	Advanced energy storage technology
aFRR	Automatic frequency restoration reserve
AI	Artificial intelligence
AMI	Advanced metering infrastructure
ANFIS	Adaptive neuro-fuzzy inference system
ANN	Artificial neural network
AR	Autoregression
ARF	Additional registration fee
ARFIMA	Autoregressive fractional integral moving average
ARIMA	Autoregressive integrated moving average
AV	Autonomous vehicle
BDA	Big data analysis
BP	Backpropagation
BPNN	Backpropagation neural network
CACS	Comprehensive automobile control system
CEVS	Carbon Emissions-based Vehicle Scheme
CFD	Computational fluid dynamic
CHP	Combined heat and power
CIM	Compute-in-memory
CNN	Convolutional neural network
CPD	Custom power device
CS	Cybersecurity
CSC	Common system consensus
CVPP	Community-based virtual power plant
DBMS	Database management system
DBN	Deep belief network
DDC	Distribution dispatching centers

DDOS	Distributed denial of service
DER	Distributed energy resource
DES	Distributed energy source
DFD	Deterministic frequency deviation
DG	Distributed generation
DG-CONNECT	Directorate-General for Communication Networks, Content and Technology
DL	Deep learning
DNN	Deep neural network
DSL	Digital subscriber line
DSM	Demand side management
DSO	Distribution system operator
DSRC	Dedicated short range communication
DSTATCOM	Distribution static compensator
DVR	Dynamic voltage restorer
EaaS	Energy as a Service
EC	Electrochemical capacitor
EC	European commission
ECF	Entire consumption of fuel
EIA	Energy information administration
ELM	Ensemble learning method
EMS	Energy management system
EPS	Electric power supply
ERGS	Electronic route guidance system
ESD	Energy storage device
ESS	Energy storage system
EV	Electric vehicle
FAN	Field area network
FAT	Full activation time
FCR	Frequency containment reserve
FCRBM	Boltzmann machines with factored conditional restrictions
FDI	False data injection
FDIA	False data injection attack
FHWA	Federal highway administration
FIR	Finite impulse response
FLANN	Functional artificial neural network
FRR	Frequency restoration reserve
GA	Genetic algorithm

GBM	Gradient boosting machine
GBRT	Gradient boosting regression trees
GHO	Global Health Observatory
GNB	Gaussian naive Bayes
GPR	Gaussian process regression
GPS	Global positioning systems
GRNN	Generalized regression neural network
GRU	Gated recurrent unit
GSDR	Global Sustainable Development Report
GVR	Green Vehicle Rebate
GWDO	Genetics in wind-driven optimization
HAN	Home area network
HC	Hysteresis controller
HVAC	High Voltage AC
I2V	Infrastructure to vehicle
ICCT	International council on clean transportation
ICT	Information and communication technology
IDM	Intelligent decision-making
IGBT	Insulated gate bipolar transistor
IIR	Infinite-duration impulse response
IOE	Internet of Energy
IP	Internet protocol
IPFS	InterPlanetary File System
ISE	Integral square error
ITLC	Intelligent traffic light control
ITS	Intelligent transport system
KNN	K nearest neighbors
L-FLANN	Legendre polynomial-based FLANN
LIBS	Lithium$-$ion battery
LMS	Least mean square
LPF	Low-pass filter
LR	Linear regression
LSTF	Linear support transfer function
LSTM	Long short-term memory
LTA	Land Transport Authority
MA	Moving average
MAAS	Mobility as a service
MAE	Mean absolute error
MAPE	Mean absolute percentage error

MARS	Multivariate adaptive regression spline
ME	Mean error
mFRR	Manual frequency restoration reserve
MG	Microgrid
MIMO	Many inputs, many outputs
MIROS	Malaysian institute of road safety research
ML	Machine learning
MLP	Multilayer Perceptron
MPE	Mean percentage error
MSE	Mean squared error
NAN	Neighbor area network
NCES	Non-conventional energy source
NLP	Natural language processing
NN	Neural network
NVDLA	NVIDIA® Deep learning accelerator
OBU	On-board unit
OSI	Open systems interconnection
OTC	Open traffic control
P2P	Peer-to-peer
PACF	Partial auto-correlation function
PCA	Principal component analysis
PCC	Point of common coupling
PET	Power electronic technology
PEV	Plug-in electric vehicle
PI	Proportional integral
PMU	Phasor measurement unit
PQ	Power quality
PRS	Prosumer recommender system
QoS	Quality of service
RE	Renewable energy
RISC-V	Reduced instruction set computer five
RL	Reinforcement learning
RMSE	Root mean squared error
RNN	Recurrent neural network
ROI	Return on investment
RTU	Remote terminal unit
SCADA	Supervisory control and data acquisition
SDAE	Stacking in denoising autoencoder
SDG	Sustainable development goals

SE	Squeeze and excitation
SOC	States of charge
SOH	States of health
SPP	Spatial pyramid pooling
SRF	Synchronous reference frame
SSP	Spatial pyramid pooling
SST	Solid-state technology
SVM	Support vector machine
TA	Time of activation
TC	Time of cycle
TCP	Transmission control protocol
TERAS	
T-FLANN	Trigonometric FLANN
TSO	Transmission system operators
TVPP	Technical VPP
UN	United Nations
UPQC	Unified PQ conditioner
UT	Unit template
V2G	Vehicle-to-grid
V2I	Vehicle to road infrastructure
V2V	Vehicle to vehicle
VAR	Vector autoregression
VPN	Virtual private network
VPP	Virtual power plant
VSC	Voltage source converter
WAMS	Wide area measurement system
WAN	Wide area network
WNN	Wavelet neural network
WPI	Wholesale price index
WT	Wind turbine

1

Innovation, Opportunities, and Ongoing Challenges of AI and IOE in the Power and Energy Sector

Manjulata Badi[1], Swetha Shekarappa G[1], Sheila Mahapatra[1], and Saurav Raj[2]

[1]Department of Electrical and Electronics Engineering, Alliance University, India
[2]Department of Electrical Engineering, Institute of Chemical Technology, Marathwada Campus, India
E-mail: mlbadi@gmail.com; sg.swethasg88@gmail.com; mahapatrasheila@gmail.com; sauravsonusahu@gmail.com

Abstract

With deregulation in the electrical sector, energy is being treated as a commodity. Further, with integration of renewables, smart grid technology and electric vehicle technology are a force to reckon with as a futuristic goal. Digital technology could revolutionize our current energy supply, trade, and consumption. Artificial intelligence (AI) powers the new digitalization model. A sophisticated software program that regulates integrated energy supply, demand, and renewable resources will be employed to accomplish this task. The arrival of AI will be crucial for the achievement of this goal. This research is devoted to study the AI techniques used in the energy sector. Legislation and complex business strategies must be implemented to meet the competition, as well as protect customers' privacy. Efficient implementation of new services and products in the digital energy market is possible with technological progress in information technology, AI, and data analysis. This study will provide an overview of AI projects, ambitions, new cutting-edge applications, challenges, and global roles in energy.

Keywords: AI, energy, electric vehicle, innovation, Internet of Energy (IOE).

1.1 Introduction

Artificial intelligence (AI) is one of the most disruptive technologies we are currently experiencing. AI-powered innovations are in greater demand as urban locations become smarter through community, technology, and policy. This article offers perspectives on how AI can be used to make cities smarter [1]. The review method is based on a systematic search of all previously published studies. Municipal growth is discussed, and various aspects of the economy, society, environment, and governance are examined. AI applications primarily aid in the enhancement of business efficiency, the advancement of data analytics, the provision of education, promoting energy sustainability, providing health-based solutions, optimally land usage, maintaining safety, facilitating transportation, and administering urban management [2]. There is limited research on the risks of widespread AI implementation. Unfortunately, there has not been much examination of the impacts of AI on urban and societal life [3].

It is crucial to incorporate these concepts when looking at smart cities; community, technology, and policy all have an impact on delivering productivity, innovation, liveability, well-being, sustainability, accessibility, good governance, and planning [4]. Even though more and more pieces on the topic are published, there is no research available that comprehensively analyzes the ever-growing body of research. This article provides an examination of AI assistance with smarter city development using a methodologic approach [5].

Among the potential applications of smart urban technologies, the expansion of infrastructure capacity, the creation of new services, the reduction of emissions, the engagement of citizens, the reduction of human error in decision-making, the support of sustainable development, the improvement of commercial enterprise, and city performance [6]. As of now, some of the most popular technologies being discussed in the context of smart cities include autonomous vehicles, big data, 5G, robotics, blockchain, cloud computing, 3D printing, virtual reality, and artificial intelligence [7]. These technological advancements are critical to solving the challenges associated with urbanization, but AI works in conjunction with these technologies to provide tremendous potential in addressing those problems [8]. AI is indisputably the most disruptive technology of them all.

There is a strong emphasis on sensor data, experimentation data, and knowledge-based data in industrial energy savings models that use data-driven methods [9]. This chapter shows that significant data science effort was applied to create data-driven models to ensure data quality in industrial settings [10]. The real challenge with Industry 4.0 is the communication of data and infrastructure, as opposed to the implementation of sophisticated modeling techniques [11]. An accurate and effective AI-based infrastructure for the industry was the goal of the study. With rapid 5G development, IoT standardization, AI, blockchain 3.0 utilization, as well as expert predictions, the industry is already getting closer to a world where AI is dominant [12]. Industrial energy efficiency and energy consumption have already been pointed in the right direction due to government efforts and policies [13]. This highlights a bright future for AI-based manufacturing of energy-saving systems. The chapter discusses many of the obstacles which are inherent in the transition from traditional manufacturing to AI-driven manufacturing; so it also discusses the critical role of collaboration between researchers and industry leaders in this transition [14]. For Industrial 4.0, cutting-edge technologies will be important in the current context of industrial energy savings [15]. The project depicted in the current work will save energy by standardizing and modularizing industrial data infrastructure to help companies for managing data more efficiently [16]. This work is an invaluable resource for industrial researchers and entrepreneurs who are interested in implementing new energy-saving systems. People continue to perceive some hope for governments, even when they are bogged down in political and policy muck [17]. Many policymakers, practitioners, and academicians believe that the next major turning point in the history of mankind will occur when smart urban technologies are widely implemented [18]. As a result of our technocentric approach in solving urban and environmental problems using technological means, the concept of "smart cities" has grown in popularity [19]. This grouping of metropolitan areas, referred to as the "geographies of disruption," applies digital technologics to rcvitalizc businesses, help define urban space, enhance the standard and performance of government services, and combat several urban problems [20]. Therefore, typical restructuring of the existing power structure with the incorporation of AI and IoT has laid the foundation for futuristic goals which has emerged as an upcoming technology called as Internet of Energy (IOE) [21].

1.2 Recent Advances in Smart Energy Industry's Adoption of Artificial Intelligence

A few of the obstacles in deploying AI in the smart energy sector includes data quality and data scarcity, AI network parameter tuning, technical infrastructure issues, a shortage of skilled professionals, integration issues, dangers, and regulatory concerns. Building energy system fault detection and diagnostics are challenging undertakings as well [22]. According to multiple research, data security and inadequate information are two of the biggest problems encountered by energy systems [23]. Poor controllers, sensors, and controlled devices affect the performance and reliability of energy system operation and data estimates. The high data dimensionality, strong coupling, and tremendous complexity of large-scale modeling of the power grid provide new problems to the energy market [24]. The complexity of the technology involved makes it challenging for grid operators to integrate wind and solar power using AI [25]. Several IT companies are actively showcasing quantum computing, one of the most effective supercomputers. Detection of cyberattacks speed up as quantum technology improves AI-based machine learning techniques and increases system processing power. Even if AI development is an effective and promising development for sustainability, implementing AI leaves a huge carbon footprint, exhibiting a direct rebound impact. Five automobiles' worth of CO_2 emissions can be attributed to a single AI learning algorithm [26]. Techniques utilizing AI rely extensively on various sorts of energy data, which increases both the direct and indirect global carbon footprints [27].

Other significant AI challenges in the energy sector include:

✓ AI's growth in the energy sector has been slowed down by a dearth of key AI skills among decision-makers, according to non-theoretical history. There is a technological gap in most organizations in comprehending the benefits of AI applications.

✓ Practical inexperience: There are many professionals with a thorough understanding of technical issues and corresponding solutions. However, it is difficult to find qualified professionals who can create dependable AI-powered applications with real practical advantages. Power companies monitor and maintain data, but digitizing it with sophisticated management software is difficult. Device malfunction and unauthorized access are all risks that are connected to data loss.

✓ Energy companies stand to lose a lot if they make a mistake; so they are wary of experimenting with new approaches that lack experience.

✓ The most significant barrier toward modernizing the energy sector is an outdated power system infrastructure. The utility industry is currently trapped in a sea of data it generates, with no idea of how or when to deal with it. However, despite having more data than any other organization, the industry's data is also spread across different formats and stored locally. The industry makes a lot of money, but it is also vulnerable because of old and insecure systems.

✓ It is expensive to integrate innovative advanced energy technologies. It is difficult to find a reputable software provider. Further, creating and configuring software is time consuming. In addition, the deployment of energy technology will necessitate significant funding and resources for software development, adaptation, and control.

✓ It is crucial to remember that decentralization and supply diversification, together with new AI technologies and increased demand, provide difficult problems for global energy production, transmission, and distribution as well as load consumption.

✓ In many developing economies, particularly in low- and middle-income nations, AI is still in its infancy. Automated control and smart metering account for roughly 10% of worldwide grid investment [28].

✓ Most consumers are at risk since they do not understand AI-based services and its operation. The protections will be effective if they are included in the power systems because existing technology is far from ideal.

1.3 Internet of Energy (IOE)

Electric vehicles will progressively take the place of conventional ones as the world moves toward electrification. As energy management technology develops, homes will cease to only be energy consumers and begin to generate their power [29, 30]. The sharing of energy information continues to be led by communication and Internet of Things (IoT) technologies. We will soon be able to centralize energy data and enhance management using the cloud platform [31]. By using big data analysis, prudent judgment, and cybersecurity, the Internet of Energy (IOE), which extends from neighborhoods to cities and nations, can provide sustainable energy for the smart society. Following are the attributes of IOE.

• One of the advantages of using a variety of energy sources is that they can all be combined into one system.

- For hybrid supply to electric vehicle, digital energy system operation (i.e., P2P operation), and highly adaptable, autonomous, and self-managing systems.
- Ability to withstand adverse weather conditions. There must always be a two-way flow of power and information. Global energy system expansion and scalability are simple and easy to support.
- The use of plug-and-play technologies such as plug-in electric vehicles (PEVs), renewable energy, energy storage, and balanced loads.
- Intelligent fault warning, real-time location and isolation of faults, a well-integrated and reliable communication network, and the prevention of cyberattacks through the integration of IoT infrastructure and security measures are just a few of the things that are mentioned.
- Blockchain enables transparent, impenetrable, and secured energy systems.

To develop IOE as a clean, affordable, and dependable energy system, AI must be applied in the energy sector. Renewable energy is being integrated with big data analysis as AI-powered IOE pays attention to being an IOE enabler and supplier while using machine learning techniques to continuously focus on smart power generation, consumption, and infrastructure. This simplifies the global operation and maintenance service by digitalizing and automating it for a sustainable future.

1.4 Artificial Intelligence and IOE Role Toward Carbon Neutrality

It is predicted that by 2050, more than two-thirds of the world's population will reside in urban areas, owing to a combination of people migrating from rural to urban areas as well as population growth. A significant increase in the number of people living in cities could lead to several problems such as lack of infrastructure for medical treatment, traffic problems, high energy use, and cyberattacks.

Achieving sustainability with the sustainable development goals (SDGs) of the United Nations (UN) plays an important role in balancing the economy, society, and environment. Realizing the SDGs and becoming carbon neutral are inextricably linked, which threaten sustainable development unless appropriate mitigation and adaptation measures are taken. On the other hand, as shown in Figure 1.1, most SDGs considerably benefit from AI-powered IOE.

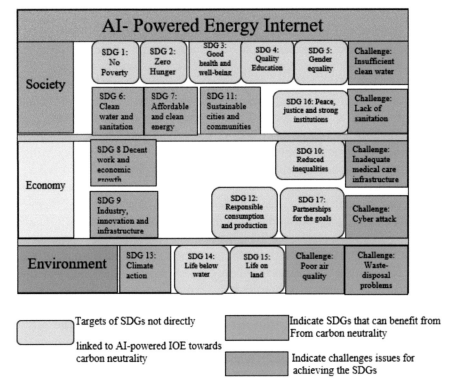

Figure 1.1 Carbon neutrality and UN SDGs connection.

Carbon neutrality necessitates the implementation of mitigation and adaptation measures at all levels. These actions can have positive synergies with SDGs or negative synergies by impeding SDGs (known as trade-offs). As a result, achieving the SDGs while also being carbon neutral requires striking a balance between synergies and trade-offs. With AI-powered IOE, an efficient solution is emerging, which can both reduce emissions and mitigate the effect of climate change while also making adaptation easier. To deal with these complex issues, we must reduce the negative effects of rapid urbanization while also implementing new technologies to deal with the inevitable consequences.

It is becoming more and more obvious that the emerging IOE concept combines various energy generation technologies and dependable energy storage technologies to develop a long-term mitigation and adaptation strategy for addressing global challenges. Our society would be substantially benefited from AI-powered IOE which would be instrumental in achieving

SDG 6 objective of clean water and sanitation, the SDG 7 goal of affordable and clean energy, and the SDG 8 goal of sustainable cities. Children and the elderly in low-income areas, as well as those with a range of chronic illnesses, are particularly susceptible to the negative impacts of environmental pollution on their health along with its social repercussions. AI-powered IOE decreases pollution by producing zero emission when giving the same amounts of energy to end user [32]. Big data analysis can be used by smart buildings to respond to cost saving, potentially postponing the need for electrical infrastructure upgrades [33]. Deep learning algorithms will enable automated renewable energy gathering across cities as self-driving EVs get more sophisticated [34]. Industry, innovation, and infrastructure will be accelerated to shift toward a low-carbon economy by generating high-quality employment and economic growth through AI-enabled IOE. The technology revolution has resulted in higher employment and faster economic growth, even though AI has the potential to replace a substantial number of occupations. When AI transforms the energy sector, more employment could be created in allied industries, particularly knowledge-intensive ones. For instance, distribution network firms require several data analyst employees to find various anomalies and boost network operating efficiency. The use of AI-powered IOE to accelerate the transition to 100% renewable energy could have a large economic benefit. Since deep learning AI improves energy efficiency and reduces emissions overall, Google's data center cooling costs have been cut by 40% [35]. AI-powered IOE cutting-edge clean technologies are designed to enable low-carbon energy systems that lead to ecologically friendly energy in our cities, and the usage of fossil fuels may be drastically decreased. This significantly lowers greenhouse gas emissions.

1.5 Motivation

Our primary focus in this section is on the survey's findings, which summarize the main obstacles to a carbon-neutral internet. As shown in Figure 1.2, the energy sector faces four major challenges in its quest for carbon neutrality.

1.5.1 IOE-digitalization

The energy system is under pressure to continue being sustainable due to the widespread integration of distributed energy resources, changes in electricity market regulation, and increased risk of cyberattacks. We can use these technologies to improve our decision-making for more complicated energy

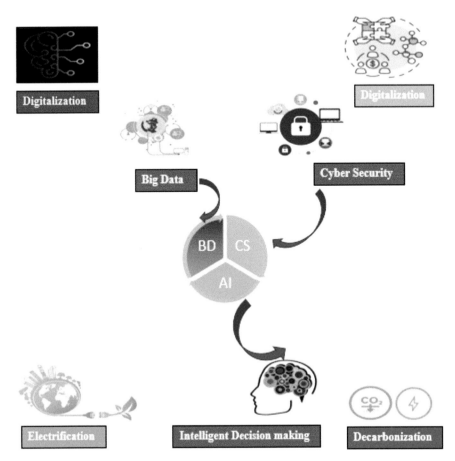

Figure 1.2 Concept of Internet of Energy toward a carbon-neutral society including (a) IOE-digitalization, (b) IOE-decentralization, (c) IOE-decarbonization, and (d) IOE-electrification.

systems. Although effectively digitalizing the energy system is a difficult task, it needs to be thoroughly researched.

1.5.2 IOE-decentralization

Energy efficiency and sustainability is increasing demand for more low-emitting power-producing infrastructures that truly decentralize the supply. As part of a green cultural shift, utilities are moving away from hierarchical management methods and toward a decentralized electrical market that permits third-party asset ownership, such as community batteries. For instance, peer-to-peer (P2P) energy trading can replace conventional electrical business models and offer new revenue for retailers and distribution network businesses. Further, decentralization would affect both supply and demand for energy.

1.5.3 IOE-decarbonization

To counteract global warming and climate change, every individual must adopt energy-efficient methods. It is intended to decarbonize large-scale, energy-intensive industrial operations, such as the nonferrous metals industry, employ heat pump for residential cooling and heating, and yield better energy efficiency with an aim to decarbonize.

1.5.4 IOE-electrification

Emergence of e-mobility has made it possible for electric vehicles to run on renewable energy sources rather than on fossil fuels. Sufficient positive movement can be achieved toward green technology by considering natural gas with hydrogen gas. However, additional investigation is required to ascertain how electrification can assist in lowering and mitigating CO_2 emissions from businesses including transportation, heating, and ancillary.

1.6 AI Application and Solution

AI applications and their solutions can be broadly classified with big data analysis (BDA), intelligent decision-making (IDM), and cybersecurity (CS). They all use artificial intelligence to handle the four concerns as outlined below.

1.6.1 Analyzing large amounts of data (BDA)

Changes in the energy sector have enabled IOE to evolve for an era of zero emissions. However, this raises a lot of challenging issues in terms of socioeconomics and technology. These advances come from a variety of sources, including smart meters and phasor measurement units that provide spatiotemporal data streams and may be understood with the help of AI, which has a wide range of IOE applications at its disposal PMUs. A wide range of important applications in the technical, social, and environmental sciences have benefited from big data analysis, and these fields are currently dealing with a large influx of data from smart meters, environmental monitoring systems, renewable energy power plants, electricity market, electrical grid monitoring systems, and physics-based models.

1.6.2 Making informed decisions (IDM)

In the practical scenario, the optimization problem objective function improves system efficiency but is frequently plagued by uncertainties. Researchers currently working on this problem are mostly using predict and optimize diagrams. Further, while employing a data-driven approach, it becomes more challenging to create a mathematical model that precisely reflects the operational process. It is possible to use deep learning and reinforcement learning to effectively solve most energy management problems [36, 37].

1.6.3 Cybersecurity (CS)

Deep reinforcement learning-based decision-making may make it simpler to choose a more effective operational approach in the face of uncertainty. Because AI algorithms are not widely known to be vulnerable, cyberattacks against the energy sector will be challenging.

Digitalization, decentralization, decarbonization, and electrification are three important problematic issues that must be addressed to transform energy production and carbon footprint.

1.7 Big Data Analysis

Big data analysis for IOE may be applied in the most cutting-edge manner using machine learning techniques including time series modeling and forecasting, supervised and unsupervised classification, fault diagnosis, anomaly detection, and computer vision. The data source that enables the replication of these findings is displayed in Figure 1.3.

Models may be developed to characterize consumption patterns, predict output using future knowledge, and estimate pricing in the face of uncertainty [38]. The AI algorithm's time series regression method attempts to deal with the considerable volatility and uncertainty of residential electricity use. The LSTM [39] architecture is a common recurrent neural network that can recognize the temporal dependency of time series data. Using publicly accessible data from home smart meters, standard LSTM-based deep networks are developed for short-term forecasting of individual residential household

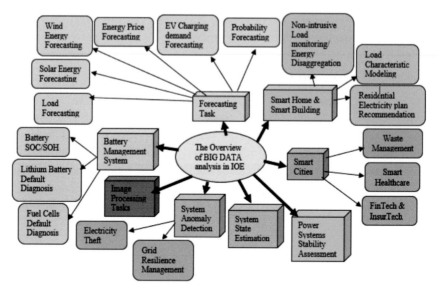

Figure 1.3   IOE-big data analysis (IOE-BDA) to achieve UN's SDGs taxonomy.

energy demand [40]–[43]. Innovative learning systems are introduced for load forecasting, which includes the extreme learning machine (ELM) and the LSTM model based on active learning [44].

1.7.1 Solar energy prognosis and prediction

Using a range of feature inputs, artificial intelligence can predict how much solar energy is to be produced by a solar farm or a rooftop PV panel [46]. For forecasting solar irradiation, a convolutional graph autoencoder [47] is created, and to take advantage of the spatio-temporal distribution of solar irradiation, a deep graph dictionary learning is built [48, 49]. Additional details on solar energy predictions are provided in [50].

1.7.2 Wind energy prognosis and prediction

Other renewable energy sources, such as wind, have more unpredictable yields. For wind energy forecasting, a residual learning with layers, which connect the multi-level residual network and DenseNet, is proposed to reduce the likelihood of the current LSTM model being overfitted. To improve wind power prediction accuracy, a generalized least squares method is introduced in [52] to reconcile wind power predictions at different levels using a hierarchical forecasting approach. For probabilistic wind power forecasting, the pair bootstrap-based ELM is used to model the regression uncertainty most effectively [53]. The supply−demand relationship is significantly impacted by the price of electricity by neural networks (NN) [54]. The maximum likelihood technique is used in the second stage to estimate the noise variance, whereas the ELM is utilized in the first stage to estimate point forecasting of price. To determine prediction intervals for market clearing prices, a two-stage approach is examined. A bootstrapping-based ELM [55] is also suggested to increase the precision of forecasting electricity price intervals.

1.7.3 Demand forecasting for electric vehicle (EV) charging stations

As the number of EVs on city streets would rise, forecasting EV charging demand is becoming more crucial for infrastructure design and administration. An ANN model is suggested, which employs Levenberg−Marquardt training to implement backpropagation learning for the job of anticipating EV charging demand. The charging need for electric buses is estimated using a longitudinal dynamics model [56] for bus transit operators with a sizable transportation network. The authors of [57] investigate a comparative case study of forecasting EV consumption using scenarios. A probabilistic model is created to represent the spatio-temporal distribution of EV charging load on a country scale for high numbers of EVs in transportation.

Gradient boosting regression trees (GBRT) were not used to perform the net load prediction job until [60]. The probabilistic forecasting procedure where customers are clustered into different groups proposes deep Bayesian LSTM networks. Quantile regression is introduced by probabilistic forecasting [59] to practically convey forecast uncertainties. Estimate theory and several statistical analytical models can be utilized for peak load estimation.

1.8 Smart Buildings and Smart Homes

1.8.1 Inconspicuous load monitoring and disaggregation

Demand response has a promising future due to non-intrusive load monitoring, which attempts to break down a single customer's power usage into individual appliance-level consumptions. The 18-layer CNN structure can classify Type-II household appliances in a better way. There is dual inclusion of the hybrid intelligent non-intrusive load monitoring approach and its properties for representation of power consumption [61]–[66].

1.8.2 Modeling of load characteristics

The learning approches may be used to partition on bigdata sets into several groups to enable demand management and energy reliability efficiency, management based on the similarity assessment that is stated [67, 68]. Customer power usage patterns are frequently compared to enhance result while forecasting and estimating residential baselines.

A household electricity plan is advised. The Australian electricity market offers a wide range of energy consumption programs, each with a distinct discount. Finding the ideal one for a certain customer is challenging. The recommendation system suggested in [69] learns about the user's experiences with various smart home appliances by utilizing a collaborative filtering technique and taking into consideration the customer's daily habits, and then suggests potential cost-saving energy strategies to them. Similar personalized recommender systems for power planning are created in [70] using a collaborative filtering technique.

1.8.3 Waste management in smart cities

Machine learning methods that are included in wastewater treatment plant modeling can be used to derive and solve an advanced energy cost function. By keeping an eye on the condition of the treatment plant with a variety of sensors to manage energy usage and discharge quality, wastewater management can be technologically driven [71].

1.8.4 Innovative methods of healthcare delivery

The development of clinical decision support systems, such as those for the detection and treatment of skin cancer, is now conceivable due to breakthrough in the field of AI [72]. IBM Watson is a fantastic illustration of an intelligent healthcare system since it can offer the best diagnosis for clinical data analysis for disease risk reduction [73].

Companies in the financial technology and insurance technology sectors are implementing AI technologies to cut costs, enhance customer satisfaction, and boost financial security. The relationship between several stocks and their temporal evolution through time can be recorded for stock prediction in [74] with the aid of temporal graph convolution. AI utilization in [75] claims analytic projects may forecast cash flow across several timeframes.

1.8.5 Evaluation of the stability of the power system

The power system's return to steady-state stability or transient stability after being disturbed is defined as the capacity of the power system to resume its initial condition, given a normal or stable condition [76]. The procedure to achieve multi-objective optimization is used for online detection under hazardous operating circumstances. An ensemble model is suggested in [77] employing a PMU data set.

1.8.6 Estimation of the current state of the system

Bayesian state estimation for unobservable distribution systems is examined as imperfect measurements stated in the article [78, 79]. A data-driven power network layout for distribution networks with medium and low voltage prediction approach is created using only historical smart meter data observed in the article [80], where statistical correlations between bus voltages are represented by a probabilistic graphical model.

1.8.7 Electrical theft detection using system anomaly observation

Theft of electricity is becoming a major issue for utilities, resulting in significant financial losses that must be identified and prevented effectively [81]. The first step in detecting energy theft is to use clustering algorithms to identify a group of customers who are highly suspected of being fraudulent. Gradient boosting machine (GBM) addresses energy theft detection, where GBM highlights feature engineering pre-processing. A demand response program that uses energy-intensive appliance detection, such as water pump operation detection, can greatly reduce the load of high-energy consumption appliances.

1.8.8 Management of the grid's resilience

Transmission line outages are frequently caused by lightning, storms, bushfires, and other natural disasters, which can cause several dependability problems. Predicting the possibility of a natural disaster is a good method to prevent disastrous effects and raise the overall resilience of the energy system.

A generic regression neural network (RNN) is used to forecast transmission line lightning outages [82].

1.8.9 Processing of images in the digital domain

The most efficient technique to increase the accuracy of image categorization is to automatically capture key image features for multiple applications using CNN-based deep learning algorithms. The PV generation forecasting problem is addressed in [83] using the ConvLSTM algorithm. Power grid assets deteriorate over time due to corrosion and aging. By doing away with the necessity for manual inspections, steel transmission tower corrosion monitoring utilizes the mask RCNN and autonomous inspection of aging electrical distribution poles which is more economical.

1.8.10 Battery monitoring and charging software

Battery safety depends on an efficient battery management system that keeps track of the battery's charge, health, and charging and discharging. The states of charge (SOC) and states of health (SOH) are relevant in this situation. A deep fully convolutional neural network is used to predict the SOC on Li-ion batteries, and the learning rate has been carefully optimized to outperform recurrent models like the linear support transfer function (LSTF) and the generalized recurrent unit (GRU). Researchers created a learning approach based on a single layer-based ELM to estimate the SOH of lithium-ion batteries in the article [84]. However, battery failures like overcharging, over-discharging, and overheating reduce the battery system's dependability and can also lead to safety issues. Battery failure can be reduced by diagnosing and eliminating faults as early as possible.

The statistical analysis fault diagnosis method includes information entropy and normal distribution [85]. Analyzing and setting reasonable thresholds for fault diagnosis will be done based on the analyzed measured signals for temperature, voltage, and current. Data-driven solutions for defect identification frequently employ machine learning and a multimodal fusion approach. Data-based algorithms are used to identify battery short-circuit issues. Battery internal short-circuit faults can be detected using the multiclass relevance vector machine. The battery's external short-circuit fault is diagnosed using a backpropagation neural network (BPNN). Battery voltage is predicted using an LSTM recurrent neural network for early fault detection.

1.9 Conclusion

Artificial intelligence (AI) has the potential to improve the efficiency of Internet of Energy, multi-energy markets, and end users. Big data analysis, intelligent decision-making, and counter measures against cyberattacks are relevant useful areas of AI applications. As a result of AI advancements, the Internet of Energy can play a more active role in achieving the United Nations' sustainable development goals. They have made it possible for a wide range of applications, from traditional energy sources to health and finance, drawing interest from both the academic and corporate communities. An in-depth look at these emerging technologies is necessary to comprehend their potential usage in various applications. The study thus provided examples of utilization of renewable energy before delving into the interpretability of probabilistic forecasting. The last sections aimed to present an in-depth analysis of a variety of research projects to demonstrate the challenges encountered by stakeholders.

References

[1] RajabRjab, Amal Ben, Sehl Mellouli, and Jacqueline Corbett. "12. AI adoption in smart cities: a barriers perspective." *Research Handbook on E-Government* (2021): 209.

[2] Huseien, Ghasan Fahim, and Kwok Wei Shah. "A Review on 5G Technology for Smart Energy Management and Smart Buildings in Singapore." *Energy and AI* (2021): 100116.

[3] Singh, Chandrani, Sunil Hanmant Khilari, and Archana Nandanan Nair. "Farming-as-a-Service (FAAS) for a Sustainable Agricultural Ecosystem in India: Design of an Innovative Farm Management System 4.0." In *Digital Transformation and Internationalization Strategies in Organizations*, pp. 85-123. IGI Global, 2022.

[4] Lim, Yirang. "Smart city for comprehensive urban management: concepts, impacts, and the South Korean experience." In *Urban Planning, Management and Governance in Emerging Economies*. Edward Elgar Publishing, 2021.

[5] Moghayedi, Alireza, Bankole Awuzie, Temitope Omotayo, Karen Le Jeune, Mark Massyn, Christiana Okobi Ekpo, Manfred Braune, and Paimaan Byron. "A Critical Success Factor Framework for Implementing Sustainable Innovative and Affordable Housing: A

Systematic Review and Bibliometric Analysis." *Buildings* 11, no. 8 (2021): 317.

[6] Gedikli, Ayfer, Cihan Yavuz Taş, and Nur Billur Taş. "Redefining Smart Cities, Urban Energy, and Green Technologies for Sustainable Development." In *Handbook of Research on Sustainable Development Goals, Climate Change, and Digitalization*, pp. 216-232. IGI Global, 2022.

[7] Testarmata, Silvia, and Mirella Ciaburri. "Other Relevant Smart Technologies: From Advanced Manufacturing Solutions to Smart Factory." In *Intellectual Capital, Smart Technologies and Digitalization*, pp. 213-224. Springer, Cham, 2021.

[8] SmileSmil, Vaclav. *Grand Transitions: How the Modern World Was Made*. Oxford University Press, 2021.

[9] Teng, Sin Yong, Michal Touš, Wei Dong Leong, Bing Shen How, Hon Loong Lam, and Vítězslav Máša. "Recent advances on industrial data-driven energy savings: Digital twins and infrastructures." *Renewable and Sustainable Energy Reviews* 135 (2021): 110208.

[10] Gökalp, Mert Onuralp, Ebru Gökalp, Kerem Kayabay, Altan Koçyiğit, and P. Erhan Eren. "Data-driven manufacturing: An assessment model for data science maturity." *Journal of Manufacturing Systems* 60 (2021): 527-546.

[11] Guerra-Zubiaga, David, Vladimir Kuts, Kashif Mahmood, Alex Bondar, Navid Nasajpour-Esfahani, and Tauno Otto. "An approach to develop a digital twin for industry 4.0 systems: manufacturing automation case studies." *International Journal of Computer Integrated Manufacturing* 34, no. 9 (2021): 933-949.

[12] Begić, Hana, and Mario Galić. "A Systematic Review of Construction 4.0 in the Context of the BIM 4.0 Premise." *Buildings* 11, no. 8 (2021): 337.

[13] Hagen, Dirk. "Sustainable Event Management: New Perspectives for the Meeting Industry Through Innovation and Digitalisation?." In *Innovations and Traditions for Sustainable Development*, pp. 259-275. Springer, Cham, 2021.

[14] Javaid, Mohd, Abid Haleem, Ravi Pratap Singh, and Rajiv Suman. "Significance of Quality 4.0 towards comprehensive enhancement in the manufacturing sector." *Sensors International* (2021): 100109.

[15] Teng, Sin Yong, Michal Touš, Wei Dong Leong, Bing Shen How, Hon Loong Lam, and Vítězslav Máša. "Recent advances on industrial

data-driven energy savings: Digital twins and infrastructures." *Renewable and Sustainable Energy Reviews* 135 (2021): 110208.

[16] Vikhorev, Konstantin, Richard Greenough, and Neil Brown. "An advanced energy management framework to promote energy awareness." *Journal of Cleaner Production* 43 (2013): 103-112.

[17] Buğra, Ayşe, Refet Gürkaynak, Çağlar Keyder, Ravi Arvind Palat, and Şevket Pamuk. "New Perspectives on Turkey roundtable on the COVID-19 pandemic: prospects for the international political economic order in the post-pandemic world." *New Perspectives on Turkey* 63 (2020): 138-167.

[18] Darby, Sarah J. "Demand response and smart technology in theory and practice: Customer experiences and system actors." *Energy Policy* 143 (2020): 111573.

[19] Desouza, Kevin C., Michael Hunter, Benoy Jacob, and Tan Yigitcanlar. "Pathways to the making of prosperous smart cities: An exploratory study on the best practice." *Journal of urban technology* 27, no. 3 (2020): 3-32.

[20] Melchor, Oscar Huerta, and Jared Gars. "Planning mobility in a fragmented metropolitan area: The case of Prague and its suburbs." (2020).

[21] Miglani, Arzoo, Neeraj Kumar, Vinay Chamola, and Sherali Zeadally. "Blockchain for Internet of Energy management: Review, solutions, and challenges." *Computer Communications* 151 (2020): 395-418.

[22] Liu, Jiangyan, Qing Zhang, Xin Li, Guannan Li, Zhongming Liu, Yi Xie, Kuining Li, and Bin Liu. "Transfer learning-based strategies for fault diagnosis in building energy systems." *Energy and Buildings* 250 (2021): 111256.

[23] Ahmad, Tanveer, and Dongdong Zhang. "Using the internet of things in smart energy systems and networks." *Sustainable Cities and Society* (2021): 102783.

[24] Ahmad, Tanveer, Dongdong Zhang, Chao Huang, Hongcai Zhang, Ningyi Dai, Yonghua Song, and Huanxin Chen. "Artificial intelligence in sustainable energy industry: Status Quo, challenges and opportunities." *Journal of Cleaner Production* (2021): 125834.

[25] Ning, Ke. "Data driven artificial intelligence techniques in renewable energy system." PhD diss., Massachusetts Institute of Technology, 2021.

[26] Zhou, Quan, Yanfei Li, Dezong Zhao, Ji Li, Huw Williams, Hongming Xu, and Fuwu Yan. "Transferable representation modelling for real-time energy management of the plug-in hybrid vehicle based on k-fold fuzzy

learning and Gaussian process regression." *Applied Energy* 305 (2022): 117853.

[27] Chen, Z., Y. Liu, M. Ye, Y. Zhang, and G. Li. "A survey on key techniques and development perspectives of equivalent consumption minimisation strategy for hybrid electric vehicles." *Renewable and Sustainable Energy Reviews* 151 (2021): 111607.

[28] World Health Organization. "Ethics and governance of artificial intelligence for health: WHO guidance." (2021).

[29] Magazzino, Cosimo, Marco Mele, and Nicolas Schneider. "A machine learning approach on the relationship among solar and wind energy production, coal consumption, GDP, and CO2 emissions." *Renewable Energy* 167 (2021): 99-115.

[30] He, Yingdong, Yuekuan Zhou, Jing Yuan, Zhengxuan Liu, Zhe Wang, and Guoqiang Zhang. "Transformation towards a carbon-neutral residential community with hydrogen economy and advanced energy management strategies." *Energy Conversion and Management* 249 (2021): 114834.

[31] Cassel, Gustavo André Setti, Vinicius Facco Rodrigues, Rodrigo da Rosa Righi, Marta Rosecler Bez, Andressa Cruz Nepomuceno, and Cristiano André da Costa. "Serverless computing for Internet of Things: A systematic literature review." *Future Generation Computer Systems* (2021).

[32] Wu, Ying, Yanpeng Wu, Josep M. Guerrero, and Juan C. Vasquez. "A comprehensive overview of framework for developing sustainable energy internet: From things-based energy network to services-based management system." *Renewable and Sustainable Energy Reviews* 150 (2021): 111409.

[33] Venegas, Felipe Gonzalez, Marc Petit, and Yannick Perez. "Active integration of electric vehicles into distribution grids: barriers and frameworks for flexibility services." *Renewable and Sustainable Energy Reviews* 145 (2021): 111060.

[34] Dibaei, Mahdi, Xi Zheng, Youhua Xia, Xiwei Xu, Alireza Jolfaei, Ali Kashif Bashir, Usman Tariq, Dongjin Yu, and Athanasios V. Vasilakos. "Investigating the prospect of leveraging blockchain and machine learning to secure vehicular networks: a survey." *IEEE Transactions on Intelligent Transportation Systems* (2021).

[35] Lee, Seongin, and Jinyoung Soh. "A Study on Mid-to Long-Term Development Directions for Energy Efficiency Management in the Age of the Fourth Industrial Revolution (1/3)."

[36] Li, Xiangjun, and Shangxing Wang. "A review on energy management, operation control and application methods for grid battery energy storage systems." *CSEE Journal of Power and Energy Systems* (2019).

[37] Qi, Xuewei, Yadan Luo, Guoyuan Wu, Kanok Boriboonsomsin, and Matthew Barth. "Deep reinforcement learning enabled self-learning control for energy efficient driving." *Transportation Research Part C: Emerging Technologies* 99 (2019): 67-81.

[38] Li, Chaojie. "AI-powered Energy Internet Towards Carbon Neutrality: Challenges and Opportunities." (2021).

[39] Sahoo, Bibhuti Bhusan, Ramakar Jha, Anshuman Singh, and Deepak Kumar. "Long short-term memory (LSTM) recurrent neural network for low-flow hydrological time series forecasting." *Acta Geophysica* 67, no. 5 (2019): 1471-1481.

[40] Poornima, S., and M. Pushpalatha. "Prediction of rainfall using intensified LSTM based recurrent neural network with weighted linear units." *Atmosphere* 10, no. 11 (2019): 668.

[41] Wang, Chongren, Dongmei Han, Qigang Liu, and Suyuan Luo. "A deep learning approach for credit scoring of peer-to-peer lending using attention mechanism LSTM." *IEEE Access* 7 (2018): 2161-2168.

[42] Jahwar, Alan Fuad, and Subhi RM Zeebaree. "A State of the Art Survey of Machine Learning Algorithms for IoT Security." *Asian Journal of Research in Computer Science* (2021): 12-34.

[43] Zhang, Chi, Sanmukh R. Kuppannagari, Rajgopal Kannan, and Viktor K. Prasanna. "Generative adversarial network for synthetic time series data generation in smart grids." In *2018 IEEE International Conference on Communications, Control, and Computing Technologies for Smart Grids (SmartGridComm)*, pp. 1-6. IEEE, 2018.

[44] Chen, Yanhua, Marius Kloft, Yi Yang, Caihong Li, and Lian Li. "Mixed kernel based extreme learning machine for electric load forecasting." *Neurocomputing* 312 (2018): 90-106.

[45] Rajadurai, Hariharan, and Usha Devi Gandhi. "A stacked ensemble learning model for intrusion detection in wireless network." *Neural computing and applications* (2020): 1-9.

[46] AlKandari, Mariam, and Imtiaz Ahmad. "Solar power generation forecasting using ensemble approach based on deep learning and statistical methods." *Applied Computing and Informatics* (2020).

[47] Dairi, Abdelkader, Fouzi Harrou, and Ying Sun. "A deep attention-driven model to forecast solar irradiance." In *2021 IEEE 19th International Conference on Industrial Informatics (INDIN)*, pp. 1-6. IEEE, 2021.

[48] Khodayar, Mahdi, Guangyi Liu, Jianhui Wang, Okyay Kaynak, and Mohammad E. Khodayar. "Spatiotemporal behind-the-meter load and pv power forecasting via deep graph dictionary learning." *IEEE transactions on neural networks and learning systems* (2020).

[49] Ahmed, Razin, V. Sreeram, Y. Mishra, and M. D. Arif. "A review and evaluation of the state-of-the-art in PV solar power forecasting: Techniques and optimization." *Renewable and Sustainable Energy Reviews* 124 (2020): 109792.

[50] Cwagenberg, Jeffrey D. "Improving Short-Term Local Solar Energy Forecasts for Optimizing Power Generation Using Machine Learning." PhD diss., The George Washington University, 2021.

[51] Sha, Huajing, Peng Xu, Meishun Lin, Chen Peng, and Qiang Dou. "Development of a multi-granularity energy forecasting toolkit for demand response baseline calculation." *Applied Energy* 289 (2021): 116652.

[52] Quan, Hao, Abbas Khosravi, Dazhi Yang, and Dipti Srinivasan. "A survey of computational intelligence techniques for wind power uncertainty quantification in smart grids." *IEEE transactions on neural networks and learning systems* 31, no. 11 (2019): 4582-4599.

[53] Ren, Chao, Rui Zhang, Yuchen Zhang, and Zhao Yang Dong. "Hybrid randomised learning-based probabilistic data-driven method for fault-induced delayed voltage recovery assessment of power systems." *IET Generation, Transmission & Distribution* 14, no. 24 (2020): 5899-5908.

[54] Li, Ranran, Xueli Chen, Tomas Balezentis, Dalia Streimikiene, and Zhiyong Niu. "Multi-step least squares support vector machine modeling approach for forecasting short-term electricity demand with application." *Neural computing and applications* 33 (2021): 301-320.

[55] Mohammadi, Shapour. "A new test for the significance of neural network inputs." *Neurocomputing* 273 (2018): 304-322.

[56] Al-Ogaili, Ali Saadon, Agileswari Ramasamy, Tengku Juhana Tengku Hashim, Ahmed N. Al-Masri, Yap Hoon, Mustafa Neamah Jebur, Renuga Verayiah, and Marayati Marsadek. "Estimation of the energy consumption of battery driven electric buses by integrating digital elevation and longitudinal dynamic models: Malaysia as a case study." *Applied Energy* 280 (2020): 115873.

[57] Zeynali, Saeed, Naghi Rostami, Ali Ahmadian, and Ali Elkamel. "Two-stage stochastic home energy management strategy considering electric vehicle and battery energy storage system: An ANN-based scenario generation methodology." *Sustainable Energy Technologies and Assessments* 39 (2020): 100722.

[58] Lingfors, David, Mahmoud Shepero, Clara Good, Jamie M. Bright, Joakim Widén, Tobias Boström, and Joakim Munkhammar. "Modelling city scale spatio-temporal solar energy generation and electric vehicle charging load." In *8th International Workshop on the Integration of Solar Power into Power Systems. Stockholm, 16-17 October, 2018.* 2018.

[59] Salinas, David, Valentin Flunkert, Jan Gasthaus, and Tim Januschowski. "DeepAR: Probabilistic forecasting with autoregressive recurrent networks." *International Journal of Forecasting* 36, no. 3 (2020): 1181-1191.

[60] Nie, Peng, Michele Roccotelli, Maria Pia Fanti, Zhengfeng Ming, and Zhiwu Li. "Prediction of home energy consumption based on gradient boosting regression tree." *Energy Reports* 7 (2021): 1246-1255.

[61] Kong, Weicong, Zhao Yang Dong, Bo Wang, Junhua Zhao, and Jie Huang. "A practical solution for non-intrusive type II load monitoring based on deep learning and post-processing." *IEEE Transactions on Smart Grid* 11, no. 1 (2019): 148-160.

[62] Zheng, Zhuang, Hainan Chen, and Xiaowei Luo. "A supervised event-based non-intrusive load monitoring for non-linear appliances." *Sustainability* 10, no. 4 (2018): 1001.

[63] Shukla, Praveen Kumar, Rahul Kumar Chaurasiya, and Shrish Verma. "Performance improvement of P300-based home appliances control classification using convolution neural network." *Biomedical Signal Processing and Control* 63 (2021): 102220.

[64] Liu, Qi, Kondwani Michael Kamoto, Xiaodong Liu, Mingxu Sun, and Nigel Linge. "Low-complexity non-intrusive load monitoring using unsupervised learning and generalized appliance models." *IEEE Transactions on Consumer Electronics* 65, no. 1 (2019): 28-37.

[65] Fan, Wen, Qing Liu, Ali Ahmadpour, and Saeed Gholami Farkoush. "Multi-objective non-intrusive load disaggregation based on appliances characteristics in smart homes." *Energy Reports* 7 (2021): 4445-4459.

[66] Wang, Mingxin, Yingnan Zheng, Binbin Wang, and Zhuofu Deng. "Household Electricity Load Forecasting Based on Multitask Convolutional Neural Network with Profile Encoding." *Mathematical Problems in Engineering* 2021 (2021).

[67] Du Toit, J., R. Davimes, A. Mohamed, K. Patel, and J. M. Nye. "Customer segmentation using unsupervised learning on daily energy load profiles." *J Adv Inform Technol* 7, no. 2 (2016).

[68] Kumar, Amresh, M. Kiran, and B. R. Prathap. "Verification and validation of mapreduce program model for parallel k-means algorithm on hadoop cluster." In *2013 Fourth International Conference on Computing, Communications and Networking Technologies (ICCCNT)*, pp. 1-8. IEEE, 2013.

[69] Varlamis, I., C. Sardianos, G. Dimitrakopoulos, A. Alsalemi, F. Bensaali, Y. Himeur, and A. Amira. "Rehab-c: recommendations for energy habits change, future generation computer systems." *Future Gener. Comput. Syst.(Accepted)* (2020): 1-41.

[70] Altulyan, May, Lina Yao, Xianzhi Wang, Chaoran Huang, Salil S. Kanhere, and Quan Z. Sheng. "A Survey on Recommender Systems for Internet of Things: Techniques, Applications and Future Directions." *The Computer Journal* (2021).

[71] Nawaz, Alam, Amarpreet Singh Arora, Dahee Yun, Choa Mun Yun, and Moonyong Lee. "Advanced predicting technique for optimal operation of wastewater treatment process: A ProActive scheduling approach." *Journal of Cleaner Production* 303 (2021): 126968.

[72] Bennett, Casey C., Thomas W. Doub, and Rebecca Selove. "EHRs connect research and practice: Where predictive modeling, artificial intelligence, and clinical decision support intersect." *Health Policy and Technology* 1, no. 2 (2012): 105-114.

[73] Tian, Shuo, Wenbo Yang, Jehane Michael Le Grange, Peng Wang, Wei Huang, and Zhewei Ye. "Smart healthcare: making medical care more intelligent." *Global Health Journal* 3, no. 3 (2019): 62-65.

[74] Badi, Manjulata. "Power quality improvement using passive shunt filter, tcr and tsc combination." PhD diss., 2012.

[75] Badi, Manjulata, Sheila Mahapatra, and Saurav Raj. "Hybrid BOA-GWO-PSO algorithm for mitigation of congestion by optimal reactive power management." *Optimal Control Applications and Methods* (2021).

[76] Badi, Manjulata, Sheila Mahapatra, Bishwajit Dey, and Saurav Raj. "A hybrid GWO-PSO technique for the solution of reactive power planning problem." *International Journal of Swarm Intelligence Research (IJSIR)* 13, no. 1 (2022): 1-30.

[77] Shiva, Chandan Kumar, Manjulata Badi, G. Swetha Shekarappa, Rohit Babu, Sheila Mahapatra, B. Vedik, and Shriram S. Rangarajan. "Solution for reactive power planning problem using salp swarm algorithm." In *AIP Conference Proceedings*, vol. 2418, no. 1, p. 040001. AIP Publishing LLC, 2022.

[78] Shiva, Chandan Kumar, Manjulata Badi, Rohit Babu, Sheila Mahapatra, B. Vedik, and Shriram S. Rangarajan. "Thyristor controlled series compensator for the solution of reactive power management problem." In *AIP Conference Proceedings*, vol. 2418, no. 1, p. 040002. AIP Publishing LLC, 2022.

[79] Swetha, Shekarappa G., Manjulata Badi, Saurav Raj, and Sheila Mahapatra. "Role of Renewable Energy Sources and Storage Units in Smart Grids." *Smart Grids and Microgrids: Technology Evolution* (2022): 147-173.

[80] Badi, Manjulata, Shekarappa G. Swetha, Sheila Mahapatra, and Saurav Raj. "A Architectural Approach to Smart Grid Technology." *Smart Grids and Microgrids: Technology Evolution* (2022): 295-323.

[81] Badi, Manjulata. "Monitoring and Controlling Water Flow Using IOT." *Recent Trends in Analog Design and Digital Devices* 4, no. 2 (2021).

[82] Biradar, Shant Kumar, and Manjulata Badi. "Analysing Efficiency and Effectiveness of Clap Switch Mechanism." *Journal of Recent Trends in Electrical Power System* 4, no. 1 (2021).

[83] Obiora, Chibuzor N., Ali N. Hasan, Ahmed Ali, and Nancy Alajarmeh. "Forecasting Hourly Solar Radiation Using Artificial Intelligence Techniques Prévision du rayonnement solaire horaire à l'aide de techniques d'intelligence artificielle." *IEEE Canadian Journal of Electrical and Computer Engineering* (2021).

[84] Liu, Hao, Jian Chen, Quan Ouyang, and Hongye Su. "A review on prognostics of proton exchange membrane fuel cells." In *2016 IEEE Vehicle Power and Propulsion Conference (VPPC)*, pp. 1-6. IEEE, 2016.

[85] AlThobiani, Faisal, and Andrew Ball. "An approach to fault diagnosis of reciprocating compressor valves using Teager–Kaiser energy operator and deep belief networks." *Expert Systems with Applications* 41, no. 9 (2014): 4113-4122.

2

Applications of Artificial Intelligence in Intelligent Combustion and Energy Storage Technologies

Kannan Chakrapani[1], Thiyagarajan Kavitha[2], Mohamed Iqubal Safa[3], Muniyegowda Kempanna[4], and Bharathi Chakrapani[5]

[1]Department of Information Technology, School of Computing, SASTRA Deemed University, India
[2]Department of Computer Science & Engineering, Koneru Lakshmaiah Education Foundation, India
[3]Department of IT, SoC, SRM Institute of Science and Technology, India
[4]Department of Computer Science & Engineering, BIT, Visvesvaraya Technological University (VTU), India
[5]School of Computer Science and Engineering (SCOPE), Vellore Institute of Technology, India
E-mail: kcp@core.sastra.edu; drtkavitha@kluniversity.in; safam@srmist.edu.in; kempannam@bit-bangalore.edu.in; bharathi.c2013@vit.ac.in

Abstract

Heat transfer is necessary for various emerging technologies, including information technology, biotechnology, nanotechnology, and low-carbon energy use. AI applications in heat transfer research are gaining traction and are being used to speed up basic research and practical progress in heat and mass transfer. Combustion is used in most energy generation systems, such as power generators and combustion engines, to convert the chemical energy stored in fossil fuels into work, turning into more usable and transportable forms, such as electricity. Combustion knowledge and technologies have been consistently updated to address the growing issues of fossil fuel depletion, energy security, environmental pollution, and climate change. Artificial intelligence (AI) cutting-edge technologies have emerged as valuable tools in various fields. Artificial intelligence (AI) applications in combustion research are gaining traction and are being used to speed up product development toward more environmentally friendly and efficient combustion systems. Solar, wind, hydro, biomass, geothermal, and other renewable energy sources are critical for creating a clean and sustainable future. These renewables are often intermittent, unpredictable, and poorly distributed physically and temporally; so directly incorporating them would significantly disrupt power networks. As a result, developing efficient and reliable energy storage technology is critical to effectively adopting renewable energy. Despite tremendous advancements in advanced energy storage technology (AEST) in the last decade, particularly for large-scale energy storage, intelligent and efficient energy storage systems are in demand more than ever. Vast amounts of data about the performance and life of energy storage devices are becoming available, thanks to the advent of the Internet of Things (IoT). Big data and advances in artificial intelligence (AI) present significant potential for optimizing and increasing AEST performance and durability and inventing game-changing solutions. The primary goal of this chapter is to provide a platform for presenting the most recent breakthroughs in the use of artificial intelligence in energy storage systems, particularly large-scale energy storage systems.

Keywords: Advanced energy storage system (AEST), artificial intelligence, artificial neural network, energy storage systems (ESS), heat transfer, renewable energy, renewable energy systems (RSS).

2.1 Introduction

With the emerging technologies of machine learning (ML), big data, the Internet of Things (IoT), cloud computing, and various other technologies combined with intelligent devices, data are generating rapidly, resulting in physical and social spaces. Artificial intelligence provides better knowledge to make proper decisions in various spaces. AI involves data collection and analysis to make better decisions based on intelligence. Machine learning and deep learning are the subdomain of artificial intelligence, which is increasingly used in various industries and research (e.g., manufacturing, automation, economy, robotics, etc.). Among the various fields, energy storage utilization is essential for the entire world.

Energy storage devices (ESD) and energy storage systems (ESS) have improved performance, reliability, durability, and intelligent managing strategies. Machine learning can effectively speed up calculations and complicated mechanisms to improve accuracy and make better decisions based on the given information as input. This chapter mainly focuses on the development, new concepts, techniques, methods, and applications of machine and deep learning in AI technologies. Also, it leverages how it is used for energy-storing devices and systems, which include hybrid ESS, battery ESS, thermal ESS, grid and micro-grid energy systems, and pumped-storage schemes.

Economic expansion has increased global energy consumption dramatically in the recent decade. According to estimates from the International Energy Agency, global energy and fossil fuel were enlarged by 19% and 9% from 2010 to 2017. These factors have improved the problem of energy scarcity and CO_2 emissions. CO_2 emissions have increased by 22% between the years 2010 and 2020. Improving energy productivity and encouraging renewable energy are becoming increasingly important. Electronics and electrified transportation are increasing the demand for mobile power sources, promoting the developing and managing of energy storage systems (ESSs) and energy storage devices (ESDs). Traditional methods and algorithms challenged the increasing complexity of ESSs and ESDs and the vast amount of front-end data. A new technique is required to overcome the tasks that traditional approaches encounter in terms of higher accuracy, efficiency, and optimization.

Deep learning (DL) and machine learning (ML) combined with big data have been successful in recent years. Integrating human knowledge into ML and DL resulted in previously unattainable functionalities and performance and improved interaction between humans and deep learning and machine learning systems by making AI understandable to human beings [1]. AI is primarily used in engineering, science, physics, chemistry, biology, materials science, data science, and computer science, such as image recognition, natural language processing (NLP), computer vision, and search engines. In addition to the disciplines mentioned above, machine and deep learning technologies have significant promise for enhancing predicting accuracy and computing efficiency in the creation and operation of energy storage devices. The contributions of this proposed study are:

  (i) Exploring AI's role in storing the energy for renewable systems.
 (ii) Analysing consumer energy consumption behaviour.
(iii) Encompassing various energy storage systems.

2.1.1 Artificial intelligence's role in energy storage

Energy storage technology is still in its early stages of development. Energy storage technologies now lack the infrastructure to comprehend and efficiently utilize energy. The market requires innovation and breakthroughs in capacity planning, long life (high battery uptime), higher ROI, and other areas for energy storage. There has never been a more critical time for companies to act quickly, optimize, and differentiate themselves in the industry or risk falling out of favor. Technology is advancing at a breakneck speed. From Chatbots to generative modeling, there have been significant advances in machine learning and deep learning. These ideas have enabled machines to process and analyze massive amounts of data.

Artificial intelligence (AI) and machine learning (ML) algorithms can help the energy storage industry tremendously [2]. AI-enabled energy storage will assist in data collection and analysis, and by employing simulations, it will provide insights into optimizing power usage and potential forecasting breakdowns.

Artificial intelligence (AI) can make standalone systems more innovative and accessible. Most crucially, AI can help harvest renewable energy sources by increasing the efficiency of power distribution, which is influenced by the end−production−consumption consumer cycle. The economic value of a renewable energy system is virtually increased when battery-intelligent storage is added. AI optimizes the system and maximizes the customer's return on renewable energy storage. It can aid in executing machine learning, predictive analytics, grid-edge computing, and big data, all of which are necessary to attain these results.

The intelligent storage will collect data on loads, power generation, weather, neighboring grid congestion, and other factors, which will be documented and analyzed in real time. AI-assisted storage can provide real-time efficient storage, increasing value for both the user and the grid [3].

2.1.2 Development of energy storage device and the system

Electrochemical ESD (includes batteries, capacitors, flow battery, and fuel cell), then physical ESD (include pressurized air, superconducting magnet energy storage, and the pumped storage and flywheel), and the thermal ESD are all examples of energy storage devices (includes latent heat storage and sensible thermal storage that is based on phase conversion material). Lead−acid, zinc−air batteries, nickel–metal hydride, sodium−ion, and lithium−ion are only a few examples of battery types. Vanadium flow, polysulfide-halide flow batteries, and zinc-based flow are examples of flow batteries [4]. Polymer membrane exchange, alkaline, solid oxides, and microbial fuel cells are some of the most often utilized fuel cells. Automobiles, portable power, consumer electronics, medical equipment, aeronautics, astronautics, and security systems all need batteries. Automobiles, consumer electronics, and energy harvesting all employ supercapacitors in applications that require high currents and repeated charge or discharge cycles. Fuel cells are utilized in vehicles, energy generation, aeronautics, and astronautics and are appealing for medium- to heavy-duty transportation. The flywheels are mainly used for an external power supply in electrical grids and in some types of vehicles to provide backup power instantly. Both the compressed air and flywheel can be used to store redundant electricity and supply it during high-demand periods. The electricity is generated using the pumped storage. Thermal ESDs are primarily utilized in buildings and industrial operations for heat storage and reuse and for solar energy storage for electricity generation. Specific energy, storage capacity, charge–discharge rate, specific power, heat sensitivity, response time, efficiency, lifetime, capital/operational cost, and maintenance are all common ESD factors.

There are five categories to the current ESD development goal: (1) cost reduction, (2) assuring user safety, (3) increasing performance and efficiency, (4) encouraging dependability and durability, and (5) lowering environmental effects. Accurate ESD modeling, which guides performance analysis and design, is required to achieve these objectives. The identification of model parameters is one of the linked issues. Appropriate ESD design, which includes structural parameter selection, material selection, and operational strategy development, is crucial to achieving the desired cost, performance, efficiency, durability, and safety. The task of optimizing the system systematically is a difficulty. The task of methodically optimizing the design for diverse settings is a difficulty.

Furthermore, monitoring and anticipating the ESD condition of a battery, such as its state of capacity and the charge, is critical for managing the method to function and adjust its controlling strategy to exploit performance, extend battery lifetime, and confirm ESD security [5]. Directly measuring and forecasting the state is frequently problematic. Distilling and obtaining relevant information from a large measurement dataset is a difficulty.

To attain maximum efficiency in energy storage and yield, proper ESS design and optimization are essential. Energy storage systems, scheduling management units, and monitoring units often originate in ESS as a grid or a microgrid. Electric ESS and thermal ESS are two examples of representative systems. Electric ESS includes battery ESS, flying wheel ESS, battery, fuel cell hybrid ESS, capacitor, hydraulic ESS, and compressed air ESS. Transportation (primarily in hybrid and pure electric vehicles), consumer electronics, grids and microgrids, buildings, and astronautic applications, among others, all use ESS. The primary goal of ESS development is to achieve high electrical distribution capability, high execution and energy consumption efficiency, durability, high-energy storage capacity, and minimal system cost. Correction and collection of a control strategy based on energy demand is a significant task for the current ESS [6].

Three types of ESS include battery energy-storing system ESS, hybrid energy-storing system ESS, and microgrid system containing energy-storing system ESS. The calculation and estimation of the design of ESS parameters, ESS position, and optimization of controlling techniques to achieve maximum ESS performance are all problems that artificial intelligence focuses on. To anticipate the multi-layer sensor neural networks, battery energy feeding in electric vehicles, and the regression tree methods, which are used in the application of DL to the energy-storing battery system (BESS). Temperature, average speed, time in traffic, distance, and ranges are used to make the prediction [7].

2.2 AI for the Development of Combustion Systems in Energy Vehicles

Artificial intelligence is a sophisticated data analysis technique used in combustion simulations. ANN is a tool used for various studies on developing internal combustion engines. The majority of this research looks at the impact of different fuel and several operating conditions on the engine's performance, emissions, noise, and other characteristics [8]. The research was conducted experimentally or using computational fluid dynamic (CFD) software. The obtained data were utilized for training the neural network (NN) based model and, as a result, to calculate specific IC engine properties. The literature shows various methods for predicting the parameters and accuracy. The current study combined the majority of the literature and presented neural network modeling applications in IC engines, as well as the methodology used to create the ANN model and its efficiency. Spark engines, compression engines, and homogeneous charge compression are the three types of engines based on various types of fuels used to describe how a charge is a compressed ignition system.

2.2.1 Artificial neural network

Computational models of the biological brains are known as neural networks. A mature human's biological brain is made up of billions of neurons. A neuron in a single adult human brain would stretch for hundreds of kilometers if stretched out end-to-end. The neurons are enormously coupled through changeable, directed linkages, although each is functionally essential. It is thought that the human brain's exceptional skills are due to its parallel distributed processing architecture. Artificial neurons provide the basis of a neural network. Like its biological counterpart, a simple computational element is an artificial neuron. It does a weighted sum of its inputs first (Figure 2.1). The activation of the neuron is the sum of these numbers. The neuron then modulates its activity using a nonlinear sigmoid transformation [9].

Weighted interconnections interconnect a multi-layer collection of neurons to form a neural network. In the INPUT layer, each input is characterized by a neuron, a neuron in the OUTPUT layer characterizes each output, and the HIDDEN layer is characterized by processing neurons with several hidden layers.

In hidden neurons, the functions are represented by an artificial neural network (ANN) to produce better accuracy [10]. A neural network starts with random weights (input) and adjusts them until it achieves the appropriate level

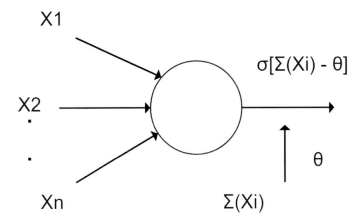

Figure 2.1 The artificial neuron.

of accuracy. The backpropagation algorithm is used to train a neural network (NN) to ensure a neural network convergence to the mapping underlying its training data [11].

The neural network is trained using the feed-forward backpropagation approach. Three layers make up feed-forward networks. In the hidden layer, there are five neurons and one output neuron. Weights from the input are used in the first layer. The weight of each following layer is derived from the previous layer. Biases can be found in all three tiers. The network output is the final layer.

The soot enabled neural network can be implemented in EMS using a polynomial that looks like this:

Soot = $X1$ + $X2$ * Fuel + $X3$ * Mass air flow + $X3$ * Mass air flow
$\qquad$ + $X4$ * Coolent temperature
$\qquad$ + $X5$ * Boost pressure
$\qquad$ + Xij (Fuel, Mass air flow)
$\qquad$ + Xij (Mass air flow, Coolent temperature)
$\qquad$ + Xij (Mass air flow, Boost pressure),

where,
$X2-X5$ describe the weight and $X1$ represents the bias of the network.

2.3 ANN-based Heat Transfer Prediction and Renewable Energy

Machine learning and deep learning are the applications of AI, where the systems can automatically learn and develop the knowledge and ability to predict with more accuracy and experience. ANN approach is to predict interface temperature. Energy and environmental challenges have gotten increasingly significant since the turn of the century. People are scrambling to find clean, efficient energy sources. Traditional fossil fuels are increasingly being phased out in favor of clean, low-carbon energy sources [12]. The energy conversion and storage processes are two significant features of energy use. Due to the fast data storage and the release of energy from renewable resources, high charges and discharge rates, as well as low cost, become a more critical requirement for the electrochemical energy-storing technology.

Fast and reversible conversion is required, particularly for energy-storing materials, including more extended durability, larger capacity, lower cost secondary battery, high-power capacitors with a high dielectric constant, and higher energy density. This promotes the growth of electric vehicles, information transfer, and other industries and confirms the efficient use of energy. However, the materials' poor energy storage capacity stifles progress. Lithium−ion batteries (LIBS) and electrochemical capacitors (ECs) are the two most crucial electrochemical energy storage systems [13]. Renewable energy is a crucial resource for future global development, and the energy sector is undergoing multiple changes that will impact its growth and resilience. In the face of climate change, conventional resource depletion, and rising pollution, the world cannot continue to waste RE's potential. Significant changes in the energy sector are being driven by the increased use of RE technologies with variable energy supply, enormous volumes of data, bidirectional energy flow, and the need to improve the use of energy storage. To fulfill the rising demand for renewable energy, production, transmission, and distribution systems must be more powerful and have more excellent process quality. An example of a neural network is illustrated as follows:

The neuron has a bias b_i and has been summed up with the weighted inputs to attain the net weight ^{net}j as given in the following equation:

$$net_j = x_i w_{ji} + x_i w_{ji} + \ldots\ldots\ldots + x_i w_{ji} + b_i. \tag{2.1}$$

The error associated with the networks' input and output is given as:

$$E = \frac{1}{2}\left[\sum_p \sum_i |t_{ip} - O_{ip}|\right], \tag{2.2}$$

Where, E is the root mean squared error value ,t_{ip} is the output of the network and O_{ip} is the desired output followed by pattern p.

Because it deals with a significant volume of data and more complicated systems, AI is critical in the energy sector. The RE industry, in particular, can be boosted by AI, thanks to improved renewable energy monitoring, maintenance, operation and storage, as well as well-timed system operations and the control [14]. Significant AI applications are included in integrating renewable energy into power systems: Grid stability and dependability, safety operations; demand and weather forecasting; effective demand-side managing; energy-storage operations; market operations and design; and greater connection between micro-grids and grid components.

2.3.1 AI in energy storage

Without appropriate storage capacity, renewable energy systems may become unstable in the face of upcoming changes in market complexity, demand swings, virtual clients, and other factors. Recent trends suggest that AI may be able to optimize them even in the absence of comprehensive long-term meteorological data. Adding intelligent storage to a renewable energy plant increases flexibility and maximizes the return on investment by allowing for fluctuating demand and renewable variable inputs because of changing climate conditions [15]. An intelligent storage system enables the establishment of intelligent RE systems, allowing RE to be used to its maximum potential. As a result, energy system providers and consumers will benefit since they will have cheaper access to energy. The growing quality of transmission storage and network technologies is a significant benefit for RE integration. An important adjusting mechanism is the ability to switch between power sources. Storage technology can help to tackle the challenges of renewable energy volatility (mainly wind or solar) and demand cyclicality.

2.3.2 Centralized control of system in AI

AI integrated into centralized control of systems will help prevent energy shortages by detecting problems early and reducing the time required to repair them. To be effective in these respects, it should have alarm systems based on reporting, statistics, a user-friendly and web-based interface, as well as a backup server for new security keys for authentication for users from many locations and other features. AI in RE is required due to the rising interconnections across grids due to vast volumes of data and information. Centralized intelligent control refers to the platforms required to regulate the supervised sites [16].

By giving ways for dealing with RE fluctuations based on experience and forecasts, AI could aid in the management of this exponentially growing data. These will help integrate RE into the energy chain, as well as more significant usage of their potential, regardless of variation. In terms of energy capacity and technology, many assets differ. They should be linked and brought to the same denominator to ensure efficiency. Changing market conditions should justify efficiency, as it does not demand the employment of a centralized intelligent unit to adapt and respond to new circumstances.

2.4 AI for Diagnostics and Numerical Tools

Deep learning (DL) is used for detection and organization of battery cells, which exhibit anomalous behaviours. To identify voltage abnormalities in storage batteries, network model is used. Defect detection, deterioration analysis, and property classifications are all things that can be done with batteries. To ensure the stability of the battery cell, machine-learning techniques are used to detect and classify faults in batteries, as well as to identify and classify anomalous batteries. This enables the BMS methods to choose proper controlling techniques for the battery safety and lifetime.

Several Machine learning algorithms, including logistic regression, k-NN, kernel-SVM, NN with only hidden layer and Gaussian nave Bayes (GNB), are used to categorize the unbalance and loss of Ni MH battery cells in both unsupervised and supervised learning approaches [17].

2.5 Next-Generation Energy Storage Technologies

We needed non-fossil fuel based alternate energy options because of the rising global temperature. Renewable energy generation is becoming increasingly attractive in the power system as it gets closer to where it is used. Because of the fast deployment of RE technologies, the power system is transitioning to a new level, necessitating adaptable energy supply, storage facilities, bidirectional current flow, and processing the vast volumes of data. According to Navigant Research, global microgrid generation capacity is expected to increase from 1.4 GW in 2014 to 7.6 GW by 2023 [18]. Power system operators face a new problem in maintaining power quality and dependability due to their alternating behavior and partial storage capacity [19].

2.6 Advanced Control Systems for Energy Storage

As energy storage technology improves, it is increasingly easier to harness renewable energy. Because the sources are intermittent, AI can help with the difficulty of capturing the distinctive combination of renewable energy production phases. The situation's complexity stems from the complex and dynamic structure of the production−consumption system. This makes real-time data analysis and monitoring complicated [19]. Having adequate control of energy storage via intelligent platforms is the key to unlocking the value. Combining renewable energy with artificial-intelligence-driven storage can be a game-changer for brighter development.

2.7 AI Applications in Power Sectors

Fault detection, coupled with real-time maintenance and the discovery of optimal maintenance plans, has been one of the most critical applications of AI in the energy sector. In a field where equipment failure is widespread and can have profound implications, AI paired with proper sensors can effectively detect and monitor faults before they occur, saving resources, time, money, and lives. Geothermal energy, which produces a consistent amount of energy, is considered a possible power source to help and support the spread of less dependable renewables (Figure 2.2).

Toshiba ESSs have been investigating the use of the Internet of Things and AI to improve the efficiency and dependability of geothermal control power plants. For instance, predictive diagnostics provided by rich data are used to predict problems that can cause plants to shut down. IoT and artificial intelligence optimize preventive measures such as chemical agent sprays (amount, composition, and timing) to avoid turbine shutdowns. Such developments are crucial in a country like Japan, which has the world's third-largest geothermal resource, particularly as the costs of competing renewable energy sources like solar power continue to decline.

Making energy-efficient decisions, customers can communicate with their thermostats and other control systems using smart devices like Google Home, Amazon Alexa, and Google Nest to control their energy consumption. Automatic meters can employ AI to optimize energy use and storage due to the digital conversion of home-based energy management and electronic machines.

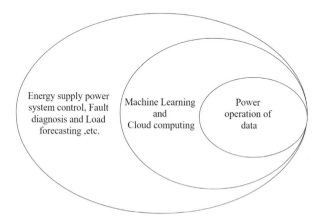

Figure 2.2   AI application in power grid.

2.8 Smart Grid with Energy Storage

With AI as the brain, intelligent energy storage seems to be one of the most essential sustainable and dependable solutions to address the various difficulties that arise. This smart grid with an energy storage system will collect and analyze the volume of data from lots of smart sensors to make appropriate decisions on efficiently distributing energy resources. As a result, we will witness an increase in reliable microgrids that manage limited energy needs with more precision. These are paired with various innovative energy storage technologies to allow for continued exchange between local settlements, even when the broader power supply is disrupted by bad weather or other breakdowns [18].

2.9 AI in Smart Grid

The foundation of intelligent and innovative energy systems is the application of AI to replace human activities to achieve high efficiency, dependability,

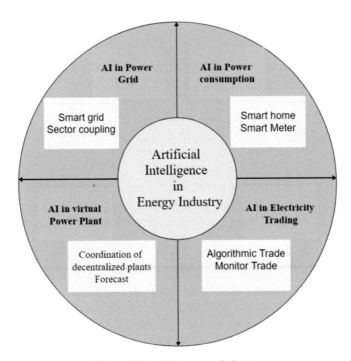

Figure 2.3   AI in energy industry.

and low cost (Figure 2.3). There are AI application opportunities at every link of the power system, including distribution, transmission, power generation, consumption, and transformation [20].

2.10 Renewable Energy Forecast in Power Generation

RE power generation is gaining popularity, but its alternating and volatile nature will influence system stability. The power system's steady, efficient, and cost-effective operation depends on the precise prediction of renewable energy generation. The classic shallow model prediction method performs poorly when dealing with non-stationary and nonlinear wind or light data. The LSTM model, based on the load prediction concept, can also be used to estimate the power generated by wind and solar panels.

2.11 Fault Diagnosis in Power System

Flexible power apparatus is made up of power electronics-based equipment. It offers services in alternating current transmission, microgrids, power storage direct current transmission, power distribution systems, renewable energy generation, and other disciplines. Fault diagnosis, detection, and protection of flexible apparatus in power systems are the first line of defense in guaranteeing equipment safety, and it is crucial in detecting faults quickly, avoiding equipment damage and reducing fault extension [19].

Its changing structure, strong coupling, imprecise control variables, and other factors that make fault diagnosis complex influence the fault characteristics of flexible equipment in power systems. Deep learning can acquire in-depth features of fault patterns of flexible equipment and transfer new information into the sample space through machine learning, allowing fault characteristics of flexible equipment to be expressed clearly at different levels.

The fault diagnosis of flexible apparatus in power systems is influenced by its changing structures, strong coupling and inaccurate control variables, and other aspects that make fault identification difficult. Deep learning can learn extra features of fault patterns in flexible tools and use machine learning to transfer new knowledge into the sample space, allowing fault characteristics in flexible equipment to be described clearly at various levels.

2.12 Analysis of Consumer Energy Consumption Behavior

Machine learning clustering and recognition abilities in AI may be used to assess user power consumption behavior, perform non-invasive load monitoring, detect abnormal power usage, and the theoretical support for reasonable energy power system pricing and energy system enhancement, as well as flexible two-way communication between energy supply and customers, comes from these analyses and tests.

For instance, AI clustering, collecting, and data analysis can be used to identify the features of different user groups' electricity ingesting behavior, achieve scientific customer segmentation, and then provide custom-made marketing and services based on data from smart meters that measure power, voltage, and current. Non-malicious variables can affect the accuracy of anomalous behavior identification results by altering power ingestion patterns over long and short time scales. As a result, it is critical to recognize and rule out the influence of these non-malicious characteristics. The terms "power consumption analysis" and "abnormal behavior detection" describe user characteristics that can be quantified using feature extraction or classification. A multi-hidden layered deep learning network can improve the performance of a classifier [19].

The application requirements, such as the application parameters shown in Fig. 2.4, must be considered when choosing a suitable form of energy storage. This will help expand the use of energy storage technologies in energy production, transportation, heating, and cooling. Mechanical power

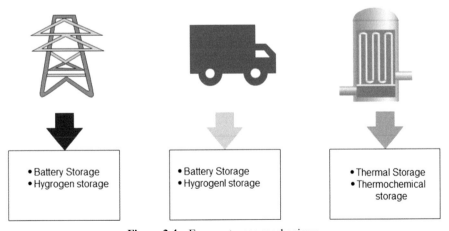

Figure 2.4 Energy storage mechanisms.

storage, electromagnetic frequency storage, electrochemical device saving, and thermonuclear energy storage are the four types of ESSs. Electrochemical energy storage has a response time of milliseconds, rated power of megawatts, and the best cycle efficiency (greater than 80%); nevertheless, the service life is limited. Portable devices, vehicles, and fixed energy resources are examples of EES applications. Energy storage and air compressor ESSs have a long lifespan and big-rated power, but they also have lengthy comment times and tight site selection requirements, necessitating large-scale promotion in high-electricity-load locations. The cycle efficiency is improved to 60% by combining battery packs, heating systems, and other technologies. As a result, we should fully use its benefits to achieve thermodynamic connection of the power source, peak trimming of the power grid, and secure the power grid's safe and stable functioning.

2.13 Network Security Protection in Power System

Information services, dynamic control, and real-time perception are part of the smart grid scheme. Because of the deep information flow interaction, the power system will be exposed to more hazards. The network assault on the power system is well-hidden and takes a long time to develop. Even if the primary equipment is not harmed directly, the secondary system can be attacked to target the physical power grid system. Deep learning can identify and detect malware, threats, and intrusion by automatically detecting network attack features, and it provides network security protection for all power systems. Anomaly sample data from an attacked power network is much lower than average sample data because the chances of a power system being attacked are much lower than in regular operation [20]. The deep learning technique does not need sample labels during the training phase, which can help to reduce the effects of a small sample size.

2.14 Conclusion

This chapter covers the use of artificial intelligence technologies and applications in the creation and administration of energy storing devices (ESD) and energy storing systems (ESS) in detail. Machine learning (ML) has shown promise in various applications, including energy consumption prediction, state estimation, fault and defect diagnostics, load forecasting, property and performance analysis, designs, modeling, and optimization. Smart grids, intelligent sensors, and the Internet of Things (IoT), posing new challenges and opportunities for energy storing systems, have enabled big data. ML models have increased prominence to utilize better big data for better decision-making and more efficient models. According to the ML models in energy storage systems, the most common DL algorithms for model generation are SVM, ANN, ANFIS, MLP, WNN, deep learning, decision trees, and sophisticated hybrid ML models. As a result, machine learning (ML) models have become crucial in such energy systems. The DL models' robustness has been demonstrated in the context of these specific applications. Due to the findings, the study focused on customized machine learning models designed for a specific application. This trend has demonstrated that case-based deep learning model creation may attain the highest accuracy level for various applications. Underneath the application conditions of a wayside energy storage system, formulate the improvement of a single train commuting strategic plan, incorporate enhancement of inter-station operational hours and division, and improve multi-train procedure diagrams using a stratified power optimization method for transmitting control coupled with energy storage. Finally, AI has aided energy storage technology in practically assisting the power grid in engineering applications, and the maturation of energy storage devices has strengthened standard specification systems.

References

[1] Çay Y, Korkmaz I, Çiçek A, Kara F (2013) Prediction of engine performance and exhaust emissions for gasoline and methanol using artifcial neural network. Energy 50:177–186.

[2] Livingston, O. V.; Pulsipher, T. C.; Anderson, D. M.; Vlachokostas, A.; Wang, N. An analysis of utility meter data aggregation and tenant privacy to support energy use disclosure in commercial buildings. Energy 2018, 159, 302–309

[3] Mohebali, B.; Tahmassebi, A.; Meyer-Baese, A.; Gandomi, A. H. Probabilistic neural networks: A brief overview of theory, implementation, and application. In Handbook of Probabilistic Models; Elsevier: Amsterdam, the Netherlands, 2020; pp. 347–367

[4] Najafi, B.; Faizollahzadeh Ardabili, S.; Mosavi, A.; Shamshirband, S.; Rabczuk, T. An Intelligent Artificial Neural Network-Response Surface Methodology Method for Accessing the Optimum Biodiesel and Diesel Fuel Blending Conditions in a Diesel Engine from the Viewpoint of Exergy and Energy Analysis. Energies 2018, 11, 860

[5] Javed S, Murthy YVVS, Ulla R, Rao DP (2015) Development of ANN model for prediction of performance and emission characteristics of hydrogen dual fueled diesel engine with Jatropha Methyl Ester biodiesel blends. J Nat Gas Sci Eng 26:549–557

[6] Ağbulut Ü, Sarıdemir S (2018) A general view to converting fossil fuels to cleaner energy source by adding nanoparticles. Int J Ambient Energy

[7] Dogaru, D.; Dumitrache, I. Cyber Attacks of a Power Grid Analysis using a Deep Neural Network Approach. J. Contr. Eng. Appl. Inform, 2019, 21, 42–50

[8] Wang, Y.; Gu, D.; Wen, M.; Xu, J.; Li, H. Denial of Service Detection with Hybrid Fuzzy Set based Feed Forward Neural Network. In International Symposium on Neural Networks; Springer: Berlin/Heidelberg, Germany, 2010; pp. 576–585

[9] Gallagher, C. V.; Bruton, K.; Leahy, K.; O'Sullivan, D. T. The suitability of machine learning to minimise uncertainty in the measurement and verification of energy savings. Energy Build. 2018, 158, 647–655

[10] Menon, D. M.; Radhika, N. Anomaly detection in smart grid traffic data for home area network. In Proceedings of the 2016 International Conference on Circuit, Power and Computing Technologies (ICCPCT), Nagercoil, India, 18–19 March 2016; pp. 1–4.

[11] Taghavi M, Gharehghani A, Nejad FB, Mirsalim M (2019) Developing a model to predict the start of combustion in HCCI engine using ANN-GA approach. Energy Convers Manag 195:57–69

[12] Livingston, O. V.; Pulsipher, T. C.; Anderson, D. M.; Vlachokostas, A.; Wang, N. An analysis of utility meter data aggregation and tenant privacy to support energy use disclosure in commercial buildings. Energy 2018, 159, 302–309

[13] Hao, R.; Lu, T. G.; Ai, Q. Distributed online learning and dynamic robust standby dispatch for networked microgrids. Appl. Energy **2020**, 274, 115256

[14] Salcedo-Sanz, S.; Deo, R. C.; Cornejo-Bueno, L.; Camacho-Gómez, C.; Ghimire, S. An efficient neuro-evolutionary hybrid modelling mechanism for the estimation of daily global solar radiation in the Sunshine State of Australia. Appl. Energy 2018, 209, 79–94

[15] Mosavi, A.; Rituraj, R.; Varkonyi-Koczy, A. R. Review on the Usage of the Multiobjective Optimization Package of modeFrontier in the Energy Sector. In Proceedings of the International Conference on Global Research and Education, Ias, i, Romania, 25–28 September 2017; pp. 217–224

[16] Qasem, S. N.; Samadianfard, S.; Nahand, H. S.; Mosavi, A.; Shamshirband, S.; Chau, K. W. Estimating Daily Dew Point Temperature Using Machine Learning Algorithms. Water 2019, 11, 582

[17] Teo, T. T., Logenthiran, T., Woo, W. L.: "Forecasting of photovoltaic power using extreme learning machine". 2015 IEEE Innovative Smart Grid Technologies - Asia (ISGT ASIA), Nov 2015, pp. 1–6

[18] Soukht Saraee H, Taghavifar H, Jafarmadar S (2017) Experimental and numerical consideration of the efect of CeO2 nanoparticles on diesel engine performance and exhaust emission with the aid of artifcial neural network. Appl Therm Eng 113:663–672

[19] Arat, H.; Arslan, O. Optimization of district heating system aided by geothermal heat pump: A novel multistage with multilevel ANN modelling. Appl. Therm. Eng. 2017, 111, 608–623.

[20] Mohammadi, K.; Shamshirband, S.; Kamsin, A.; Lai, P.; Mansor, Z. Identifying the most significant input parameters for predicting global solar radiation using an ANFIS selection procedure. Renew. Sustain. Energy Rev. 2016, 63, 423–434

3

Sustainable Smart Energy Systems and Energy Preservation Strategies in Intelligent Transportation Sectors

Bharathi Chakrapani[1], Kannan Chakrapani[2], Thiyagarajan Kavitha[3], Mohamed Iqubal Safa[4], and Muniyegowda Kempanna[5]

[1]School of Computer Science and Engineering (SCOPE), Vellore Institute of Technology, Chennai Campus, India
[2]School of Computing, SASTRA Deemed University, India
[3]Department of Computer Science & Engineering, Koneru Lakshmaiah Education Foundation, Greenfields, India
[4]Department of IT, SoC, SRM Institute of Science and Technology, India
[5]Computer Science & Engineering, BIT, Visvesvaraya Technological University (VTU), India
E-mail: bharathi.c2013@vit.ac.in; kcp@core.sastra.edu; drtkavitha@kluniversity.in; safam@srmist.edu.in; kempannam@bit-bangalore.edu.in

Abstract

In the context of the digital world, the upcoming technologies in the transportation sector electrify humans for a safe, eco-friendly, and many suitable means of journey in each dimension. Intelligent transportation system modifies transportation at basic and advanced levels by integrating the sensors and suitable PCs inside the vehicles across the travel ecosystem. Today's transport sector employs a traffic management system that consists of collecting information, transmitting information, and examining data and details about travelers for streaming real-time traffic data. Moreover, driverless auto, cabs, castor, modern subways, hyperloop, and flying taxis are some of the upcoming technologies that will be uprising the globe. Mobility as a service (MAAS) is an arising system that provides bespoke offers, including car and bus hire, quicker payment, and planning of the end-to-end trip. From the traveler's eye, intelligent transport systems (ITS) present a global range of user services to secure toll payment, intelligent parking services, and route guidance. Many ITS systems, including optimizing trips, choice per trip, and lowering dependence on exported oil, pave the route for enhancing air quality and consuming energy in transportation systems. Eco-driving and eco-routing are excellent sustainable techniques for supporting drivers to save fuel usage. This also helps the driving style by delivering real-time feedback and road preferences. Despite numerous advantages of the new energy-efficient technology in ITS, some obstacles hinder performance. These include the growing urbanization, invasion of developing vehicles, shortage of energy storing devices, and enhanced greenhouse gas emissions.

This chapter aims to look at innovative ideas for green transportation, emphasizing investigating sustainable practices in ITS, designing environmentally friendly vehicles to reduce greenhouse gas emissions, and developing vehicle control technology to avoid crashes.

Keywords: Eco-driving and eco-routing, green transportation, intelligent transport systems, mobility as a service (MAAS), vehicle control technology.

3.1 Introduction

The considerable growth of people in urbanized cities due to rapid development in industrialized and emerging countries generates an increase in population density and car ownership. The total number of registered cars in the world climbed by 530.61%, from 98 million in 1960 to 618 million in 2009 [1], indicating that the total number of vehicles in the world increased by 530.61% in the last half-century. Another example is Beijing, where vehicles doubled in eight years, from 1.578 million in 2000 to 3.504 million in 2008 [2]. Traffic congestion is caused by increasing the number of cars and insufficient transportation infrastructure expansion [3]. For example, Beijing's average motor vehicle inventory growth rate is 10.91%, but the average urban road length growth rate is just 3.64% [4]. The disparity between vehicle and road capacity growth rates will provide transportation access and mobility dilemma. Expanding road capacity to alleviate traffic congestion caused by increased vehicle numbers is not the most excellent option. Congestion on the roads increases carbon dioxide (CO_2) emissions, generating pollution that impacts the global climate. This is because vehicles stuck in traffic travel at a slower speed and stop more frequently, resulting in increased fuel consumption [7].

Increased gasoline usage has negative environmental and economic consequences. Between 1997 and 2002, total CO_2 emissions from automobiles in China were predicted to have increased by 55% [8]. According to [9], the transportation sector consumes 29.5% of global energy. It was also discovered that the United States (US) used the most energy in the transportation sector (40%) rather than in industry, agriculture, commerce, or civilization. Petroleum and diesel are non-renewable energy sources, making transportation's long-term viability questionable. Congestion is not an isolated problem because it is strongly linked to public access and movement, with many consequences, including economic, safety, and environmental deterioration. As a result, boosting sustainable access and mobility through intelligent transportation systems (ITS) is one of the options to alleviate the pressure of transportation. ITS is becoming a hotbed of study among automakers and academics. Over the last few years, numerous studies have been conducted to learn about ITS [10–15]. ITS can assist in attaining sustainable mobility and, as a result, can enhance energy efficiency. Telematic [16] combines surveillance technology with communication and information technologies. ITS [17] refers to applying these technologies in transportation. The creation of ITS as a state-of-the-art method based on numerous technologies has improved the transportation system's performance through intellectualization. Different performances are possible when appropriate technologies are used in various ITS applications. Some of the most considerable ITS applications under these strategies are management systems based on traffic control, information systems based on traveling, public transportation automations, commercial vehicle systems, and vehicle control automation [17–20].

3.1.1 Power consumption in data centers

For the past few decades, ITS has been implemented in several countries. In the realm of ITS, each country has its programmers, research groups, organizations, plans, or methods to improve transportation performance. The United States pioneered the ITS concept, launching the electronic route guidance system (ERGS) in 1967. However, due to financing difficulties, the Federal Highway Administration's (FHWA) study project failed in 1971 [21]. The cancelation of ERGS research had stifled ITS research and development in the United States. As a result, little progress was made throughout the 1970s and 1980s. On the other side, Japan had been sparked by the United

States. Japan saw the value and importance of ITS and its growth potential. The anticipated ITS benefit from Japan's perspective prompted them to take the initiative to become an ITS pioneering country in 1976 by establishing a navigation system known as the comprehensive automobile control system (CACS) [19], which had proven the practicality of ERGS technology. CACS had brought ITS to the notice of other countries. Due to cheaper and more reliable communication technology, CACS arises from the communication system, advanced traffic mobile information, and communication system based on automobiles [19].

Meanwhile, CACS inspired the private sector in Germany to create the ALI-SCOUT route guidance and information system for drivers in 1979 [21] and promoted the resumption of ITS in the United States in the 1980s. The start of ITS in three Asian countries, Malaysia, Singapore, and South Korea, is then examined. Since the 1990s, these three countries have been developing ITS. Despite being new to ITS compared to the United States, Japan, Singapore, and South Korea, they have been able to develop impressive ITSs that are competitive with the pioneers and even outperform them by putting great effort into ITS development with an emphasis on electronic toll systems and public transportation systems. As a developed country, Singapore has a body called the Land Transport Authority (LTA) that is in charge of ITS development via i-transport, which integrates all available ITSs. Because of the tiny geographic area, the transportation system's management and development plan will be more consistent and efficient.

The first master plan of South Korea is a collection of ITS standards that act as a guideline to keep its ITS development on track. Meanwhile, the ITS master plan 21 is a 20-year framework for ITS, which demonstrates South Korea's commitment and initiative in realizing the aim of ITS. Being a developing country, Malaysia is a little behind the times in terms of ITS technology. On the other hand, Malaysia takes the lead in going forward in ITS by investing more in research and development to keep up with the latest ITS technologies. Malaysia has TERAS, responsible for internal ITS research and development, and MIROS, responsible for road safety research. Apart from the countries mentioned, many more are beginning to invest effort into installing ITS to create a sustainable transportation system. Because ITS provides multiple benefits worldwide, the ITS World Congress was established in 1994 as an international platform for knowledge transfer and exchange in the field. This event is held annually in three major geographic regions (America, Europe, and the Asia Pacific) and is organized by ITS-America, ERTICO, and ITS Asia Pacific [22].

3.1.2 Advantages of ITS

The following are the major benefits of an intelligent transport system [23].

   i. Advantages in terms of health, safety, and the environment

Traffic-free zones and low-emission zones are being established in cities to prevent pollution and premature mortality. By combining car systems with mobile communications and intelligent mapping technologies, the UK may save up to 14% on gasoline, or 2.9 million barrels per year. "Smart highways" increase traffic safety and capacity: casualty reductions, traffic congestion reductions, and enhanced environmental factors are all tangible benefits of average speed cameras.

   ii. The advantages of using public transportation

By using electric and hybrid automobiles, pollution is getting eliminated. Travelers use online information regarding automobiles and obtain more knowledge. Moreover, the administration data enhances the system, and electronic ticketing permits quicker and easier public transportation levels.

   iii. Benefits to drivers and traffic management

Driver information is available on the cloud and handheld devices. The equipped automobiles are constructed to enhance reliability and secure road travel using driver assistance devices.
The contributions of this research are:

   i. Exploring energy preservation strategies and resilient smart energy systems and its communication methods for a sustainable environment.
   ii. Encompassing green communication for intelligent traffic management and its role in pollution prevention.
   iii. Investigating various techniques for reducing greenhouse gas emissions through energy management with long-term mobility.

3.2 Sustainable Smart Energy Systems and Energy Preservation Strategies in ITS

The vital resource of greenhouse gases prevailed in the transportation sector worldwide. Scientists are working hard to neglect those emissions, converting fossil fuels to alternative fuels to increase traffic flow and reduce traffic congestion and greenhouse gas emissions. The few pollutants emitted by automobiles are carbon dioxide, carbon monoxide, nitrogen oxide, and particulate matter. So, the intelligent transport system is introduced to lower these emission and consumption problems. To give a cutting-edge green solution, the ITS application environment is emphasized. In order to obtain the route which utilizes fewer resources, only some beliefs must be taken to reach the standards. The car has a bunch of equipment, including global positioning systems (GPS), on-board units (OBU), road maps, etc. The vehicle-to-vehicle connection is established using the dedicated short range communication (DSRC) methodology. So, the drivers in each car drive according to traffic flow characteristics. The fuel-consuming navigation method calculates the optimum green path and provides the fuel-efficient route to the driver. While a driver plans to travel to a specific location, ITS can transfer the query to the navigation server with the help of a particular vehicle location and corresponding destination. The most appropriate route must be decided to reach our destination by employing current and historical traffic data. The structured manner of driving advisor should meet the following goals and requirements:

- The real-time traffic data are collected and updated on the intended path using a green navigation system.
- The flow rate of vehicles is computed correctly, if the traffic flow theory is used.
- The density of automobiles at a particular time is computed using the old traffic data.
- The average green speed is attempted to be maintained (40–90 km/h) as low as possible to maximize fuel efficiency and reduce pollution.
- The dynamic speed limit should be designed to meet the objectives and necessitate green driving.

The green navigation system investigates different efficient routes and chooses the route that consumes a low amount of fuel. The system avoids manually collecting traffic signals and tolls. It does not select a route to a place where the traffic may be high. The most fuel-efficient way to start at a place is shorter or sometimes more protracted. Four aspects change the condition of fuel usage in city streets. They are i) parameter's respect for the static street, ii) parameter's respect for the vibrant street, and iii) parameter's respect for the specified car. As the name implies, static parameter does not modify the behavior of streets in the city.

For example, roadway speed limits are rarely changed, and the entire traffic signal on the street is relatively constant. Characteristics that change over time are known as dynamic street parameters; for example, the medium speed on a roadway or the amount of traffic on the street. The fuel economy of a street is determined by the combination of static and dynamic roadway factors. Other factors influencing fuel usage include the vehicle type and the individual's driving style. For instance, compared to a large car, a small car considerably produces a large amount of fuel. On the other hand, fuel consumption also increases or decreases depending on the driver's mindset. On the urban traffic streets, linear type is the method that predicts the consumption of energy at the peak level of accuracy. Based on a dynamic parameter, the different aspects are mentioned. "SD" indicates starting point to endpoint. The speed of a particular way is indicated by "VS." VS includes VS, VS1, VS2, etc. Consider the following:

- Entire consumption of fuel [ECF] = consumption of fuel at current time + consumption of fuel at the vehicle's stop.
- The outcome of the model has been expressed in terms of ECF followed by the given equation [26].
- Where ECF= Entire Fuel Consumption, RL indicates the road's length, VS_i = average speed of road RL, CF indicates saving of fuel/ sec if the automobile is vegetated, and lj= lazy time of j.

3.3 Communication Methods in ITS

In order to communicate between vehicles, two processes are introduced. They are cooperative with Multiple Inputs, Multiple Outputs (MIMO), and a cooperative model. It can also neglect the transmission energy and upgrade the work. It can overcome the temporal and spatial gains in the wireless distributed network. The infrastructure to vehicle (I2V) and infrastructure to infrastructure also utilizes the cooperative mime. Here, there are limitations for the wireless nodes deployed in the road network. The data transmission takes place within the metropolitan network in the wireless ITS through road infrastructure to vehicle (I2V), vehicle to road infrastructure (V2I), and vehicle to vehicle (V2V). The essential thing is energy limitations for road sign infrastructure because batteries present in traffic road signs should not be changed. Generally, the infrastructure to vehicle transmission happens from medium to long distance, and at that time, it requires a higher energy level. On the other hand, the multichip routing method is also used since it is unreliable in energy consumption. Finally, the relay and cooperative MIMO are better techniques for energy efficiency.

Transmission of the message occurs between road, vehicle, and relay in relay transmission, as shown in Figure 3.1. Next, it transmits from relay road to vehicle in the signal combination. In order to neglect the consumption

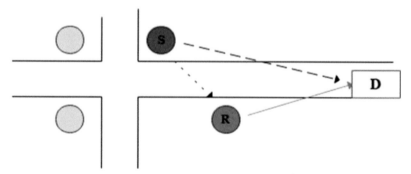

Figure 3.1   Relay transmission between vehicle and infrastructure.

of energy, the transmission diversity gain is used for a similar error rate requirement.

Another suitable technique for energy consumption is the cooperative many inputs, many outputs technique. Here, the heterogeneous reach of the MIMO spatial and temporal coding method is exploited to eliminate energy efficiency in the distributed wireless network. The best-chosen nodes depend on the topology and transmission distance to cut off the overall energy consumption.

The message transmission occurs between the *S* and *D* nodes by crossing the roads in the MIMO transmission. The thing to be understood is that the total energy consumption is computed using all the nodes. The overall energy saving of the entire network is 69% or 85% using cooperative techniques. Every cooperative transmission node equally shares the energy saving of

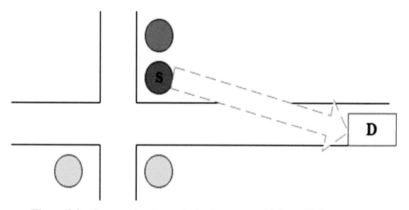

Figure 3.2   Cooperative transmission between vehicle and infrastructure.

transmission. The multichip method requires additional four hops between signal and destination nodes.

3.4 Vehicles Depending on Carbon Emissions in Intelligent Transportation Systems

In this section, the Singapore feebate method [29] is discussed. This method offers a discount to the vehicle that consumes more fuel than a vehicle with less energy. The new car or vehicle with high emissions is charged more money. The general charge for low-emission cars is between $5000 and $20,000SGD. Labeling is required to differentiate the low- and high-carbon emission automobiles. Carbon-emission-based vehicle strategy is the name of the initiative. Feebate is a policy kind. It covers new cars, taxis, and recently old foreign cars with CO_2 release of 160 or 211 g/km.

Singapore began a feebate program for low-carbon automobiles on November 28, 2012. The Carbon Emissions-based Vehicle Scheme (CEVS) took effect on January 1, 2013, replacing the Green Vehicle Rebate (GVR) program, which had run out of money on December 31, 2012. In 2001, the GVR program was established to encourage the purchase of environmentally friendly vehicles. Feebate programmers give rebates to more fuel-efficient vehicles while charging fees to less fuel-efficient vehicles. ICCT's Feebate Program Design and Implementation Best Carbon neutral vehicles furnish all necessary details for designing low-carbon automobiles.

As per the ARF, automobiles like cars, taxis, and imported vehicles with a low carbon emission level are lesser than or equal to 160 g of CO_2/km. It started on January 1, 2013, and the chargers range from $5000 to $20,000 (SGD). Low-emission taxies will get charges ranging from $7500 to $30,000. Taxis have more than 50% of charges than cars because they have high mileage and release a high amount of carbon dioxide.

Taxi companies: Since the non-Euro diesel cars release a little more delicate particulate matter, it is not applicable to the ARF funds. However, it is within the charge emission limits. Both diesel cars and taxis fall into this division.

Carbon-intensive vehicles: As per the ARF, automobiles like cars, taxis, and imported vehicles with a high carbon emission level are lesser than or equal to 211 g of CO_2/km. It started on January 1, 2013, and the chargers range from $5000 to $20,000 (SGD).

The taxis with high carbon emissions are 50% higher than the other cars and charge from $3,500 to $7,500. The new labeling idea introduced in 2012 is the fuel economy label. Under this scheme, the labels are fixed to new automobiles and light-good vehicles on display. Singapore's land transport authority manages it. The label consists of a charge amount to cheer the users to change from high-carbon vehicles to low-carbon vehicles.

3.5 Optimized Transportation for a Sustainable Environment

While the battery is essential to reduce the negative impacts, this concept is intelligent bike control [30] system (Figure 3.3) that eliminates the physical loads if it is overweight. It has three essential parts in architecture. i) cycling computer ii) motorized cycle iii) sensors. It handles the user's situation and also the surrounding of the user.

- Cycling computer: It depends on the user's setting, the amperage flow to the motor from the batteries. The indicator shows the charge percentage. It keeps track of the bicycle and detects its motion. It also restricts the speed of the cycle.
- Motorized wheel: It is the most demanding technique. It includes some parts such as a wheel, electric motor, power transmission, and brakes.
- Sensors: The sensor collects the data of bicyclists and performs the action related to it. The effects of bicycle motion are created by some parameters depending on the bicyclist's mentality and surroundings. Those parameters are detecting cyclists, timers, and sensors of the pulse. These are also needed to find the mentality of bicyclists.

As shown in Figure 3.3, the sensor's reading is saved in the database section. As we already said, the sensors are fixed in the cycle's wheel to collect the data. The parameters are also changed depending on the climate of the surrounding. Consider a situation the cyclist became ill; at that time, the cyclist switches their control to the controller. This process depends on the heartbeat rate of the cyclist. The heartbeat rate of a cyclist is as follows:

$$(220 - AC - RHR) = THRP + RHR,$$

where THR is the training heart rate of cyclists, RHR is the resting rate of heart, AC is the cyclist's age, P is the preparation level coefficient of the cyclist, $T = 0.61$ for the novice, $T = 0.66$ for the average practice of man, and $T = 0.71$ for compliant humans. Compared to THR, the cyclist has a high heartbeat rate; then, the electric drive takes charge until the heartbeat

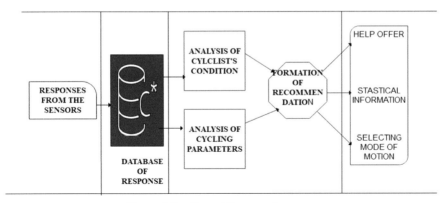

Figure 3.3   Smart bike control system.

decreases to normal. This technique is a fully independent one. It consumes more energy when compared to other strategies.

3.6 Intelligent Traffic Management using Green Communication

In order to achieve fuel consumption and exhaust pollution, a maximum number of ITS applications is utilized. The ITS-dependent technologies are summarized below. The ITS system is the best for the safety and efficiency of road traffic. This system's motto is to cut off the time of traffic congestion queue. It decreases the wait time at signals of traffic control. It employs wireless communication between the RSU and the vehicle. It has the following three effects: (a) it reduces congestion; (b) it has an economic pollutant, and c) it reduces pollution. The following vehicle runs in a pattern of stops and goes, utilizes high fuel, and emits higher pollutants than the vehicle that moves at a general speed. The vehicle does not travel much and follows the stop-and-go pattern, considered as much average speed. Then the result is that the emission rates per mile are relatively high. If the car's engine is operating yet not moving, the emission rate/mile is infinite.

To cut off the pollutant emission caused by continuous driving, Wen [34] introduced a three-tier dynamic OTC method design. Traditional methods show that the ITLC system is based on the assumption that each car is outfitted with GPS, an on-board unit (OBU), and a navigation system. GPS devices gather data about the car and its location on the road. OBU device sends the data according to vehicle speed, acceleration, and direction. The algorithm used is intelligent traffic light control (ITLC), which processes all the data and reasoning. Figure 3.4 shows a quick summarization of three-tier open traffic light control (OTC) concept.

Tier 1: It is in charge of gathering traffic data, receiving light phase data, offering traffic flow data, and estimating advised speeds [32]. As already said, GPS will share the data about the vehicle's current state, and OBU will send information to ITSC. Drivers can save time and cut off the number of stops by using ITLC [33].

Tier 2: It regulates by getting and storing the traffic flow data from the OBUs and delivers the control result to ITSC. It is split into three parts: antenna, storehouse, and traffic lights. The antenna of the OBU device in tier 1 can contact other devices via wireless communication. It permits the traffic light to get real-time traffic flow data. Simultaneously, the result of traffic controls will be contacted with OBU, permitting drivers to know about the traffic light phases in real time. The storage motto is to store the information from the gathered traffic flows. The control results are presented in the traffic lights, which are considered as displays [33].

Tier 3: Data processing tasks from the three sections are completed in tier 3. Section 3.1 deals with data extraction. Open traffic control system with three

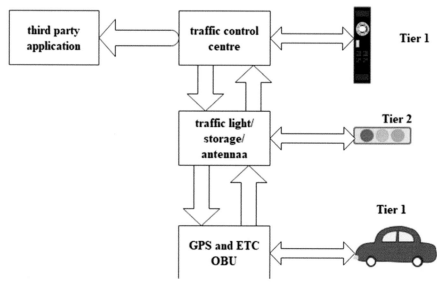

Figure 3.4 Three-tier open traffic control (OTC) system.

tiers accepts traffic data from automobiles regularly. This tier does data processing tasks and receives data from ITSC's tier 2. The ETC system collects road traffic flow data and advises the best speed.

3.7 Precaution from Pollution in Intelligent Transport Systems

This section discusses vehicle detection and counting. It includes a general vehicle classification and counter process. The process allowed for counting vehicles passing through is background registration, subtraction, and detection of foreground objects. The foremost steps include background subtraction, noise reduction, and edge recognition. In estimating highway traffic congestion, vehicle counting and detection are the toughest. The main motto of the introduced vehicle counter is to enhance the technology for automatic detection and counting of traffic vehicles on highways. The rising technology can detect and count dynamic vehicles. Figure 3.5 shows a block diagram of the proposed vehicle image counter.

a) Sequence of input frames: The traditional model uses the input frame sequence in sequential order. The mainframe is the reference frame. Then the following frames are used as input in a row. The background is registered after these frames are compared. The frame will be kept only if the vehicle is found. The detected vehicle is tracked using different techniques. If it is not found, it will be eliminated.

b) Background investigations: The algorithm used is an adaptive background subtraction algorithm. Here, the first frame itself is used as the first frame.

c) Background removal: Computer vision applications like traffic monitoring involve detecting, tracking, and counting vehicles, human−computer interaction, digital forensics, and so on. Background subtraction can be accomplished in a variety of ways. The frame differencing method is a popular method for removing the background.

In order to calculate the background image, the difference between FR1 and FR1+J is computed. When comparing the differences, the similar values of image pixels are neglected. The object is found using the learned background designing technique [35]. It occurs with the help of separating the specific feature from the pattern of the video. The difference is calculated between the first and foreground−background images [36]. So, after this, a clear image is obtained using the segmentation process. It gives a clear view of the image [37].

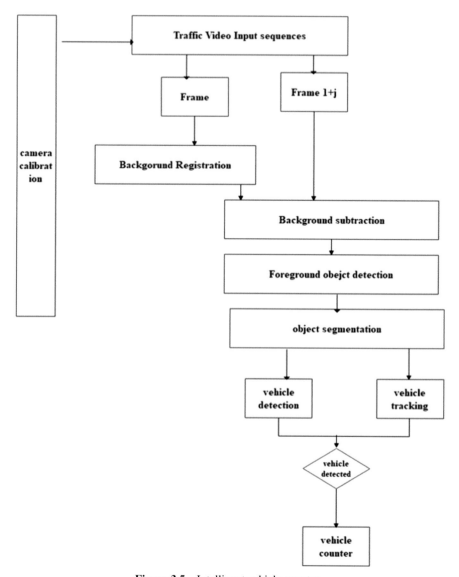

Figure 3.5   Intelligent vehicle counter.

Initially, the particular characters are opted, and automobiles evolved from that behavior. The specific features are reduced after the inclusion. It also distinguishes the input patterns from the remaining characters. The steps for vehicle counting are as follows:

1. To find the automobile and explore the foreground mask picture.
2. Appear in the county register while the registration of an automobile occurs.
3. Expand the count and register with the new label count while the automobile is not encountered.
4. Steps 2−4 are repeated until the traverse is accomplished.

3.8 Energy Management Techniques for the Reduction of Greenhouse Gas Emissions

The essential ITS projects that focus on cutting off greenhouse gas emissions are discussed in this section. It is more concerned about connected vehicles. More of these projects were funded by the following sources:

- European Commission (EC)
- European Commission Directorate-General for Communication Networks, Content and Technology (DG-CONNECT)
- European Union's seventh framework program for research (FP7)
- EU Competitiveness
- Innovation Framework program and its policy support program

The EU performed an essential program in integrated automobiles called BeCoMove. It focused on cooperative mobility systems and services for energy efficiency. The integrated result is created by comprising eco-driving support and eco-traffic management. It tackles the core source of energy waste by passenger and commercial vehicles. The Ecomotive project targeted the three-core reason for avoidable energy use by road transport to reduce fuel wastage.

- Inefficient route choice
- Inefficient driving performance
- Inefficient traffic management and control

The eCoMove project showed fuel savings of around 10%−20% for eco-driving and approximately 10% for traffic signal operations [39]. The other necessary project was the ECOSTAND [40] project. It was a collaboration of the EU-Japan-US task force on enhancing a fuel saving technology for detecting the effects of ITS on-energy efficiency and carbon dioxide emissions. In order to quantify the impacts of ITS on-energy efficiency and CO_2 emissions, the three European regions, likely Union, Japan, and the USA, agreed on a framework for general assessment technology. The next project, compass 4D, is an active project in Europe. Its duration is three years, and it concentrates more on the cooperative system's benefits [41]. CO_2 emission reductions were generally on the order of 5%–15%. From 2010 to 2013, the BCooperative Networked Concept for Emission Responsive Traffic Operations, or ConCERTO, [43] project used tough platform tools to build next-generation technology for reducing motor vehicle emissions based on real-time emission data. TThe CARBOTRAF [44] project is currently in progress to create a Decision Support System for Reducing CO2 and Black Carbon Emissions via Adaptive Traffic Management. The platform was created to examine the effects of ITS introduced by Amitran [42]. It started in 2012 and ended in 2014. The perfect method for reliable integrated vehicles to produce electric vehicles was discovered by EcoGem [45] between 2010 and 2013.

3.9 ITS for Sustainable Mobility

As per the report of Accenture in 2018 [55], the gains from the manufacturing of classic cars and sales will decline by 4 million in the year 2030. However, again the mobility services may obtain 220 billion. The main merit of mobility as a service (MaaS) is that it permits users or customers to select on-demand automobiles such as cars, buses, and bicycles from their location. Specifically, in overcrowded places, the MaaS had supported to modify the automobile business. It permits compromise from the journey preparation to billing and gathers second options from different sources.

The transportation system is being developed since ITS integrates various technologies such as sensing, processing, and other communication systems. In ITS, heterogeneous communication systems elaborate their aspects to enhance performance. The fundamental use of ITS is to construct a network of transport reliably and securely. The future ITS only focuses on road safety, traffic congestion elimination, and enhancing energy efficiency rather than anything else. The structure and deployment of communication and data technologies are called green ITS. It uses a lower amount of energy but releases lesser pollutants. The two main mottos of the traffic control system are eliminating energy consumption and enhancing traffic at the allowable level of security. 1) Improving the quality of air, 2) less energy utilization, and 3) lower noise levels are the first and foremost objectives of green ITS.

The amount of gas used is always directly proportional to the pollution produced. As per the research, the driving patterns with high acceleration and power demand, late gear modifying behavior, and indicated speed interval negatively impact energy consumption and emission. Ericsson [48] proved that highly aggressive drivers have excessive energy bills. Researchers [49] found that timely gear change and level of absolute acceleration are critical for fuel consumption reduction. Compared to non-assisted driving, the authors recommended a fuel-efficient tool that saved 16%. Brundell-Freji and Ericsson [50] presented four aspects of traffic light density, speed limit, street function, and neighborhood type that influence road network behavior. According to the practical work done by the author of [51], the road grade plays a significant role in energy utilization. The efficiency achieved is 15%−20% at the link and route levels. This relationship impacts the design of environmentally friendly routing engines, as map elevation is considered in the computations. In terms of traffic circumstances, full vehicle stops and traffic congestion impact energy usage.

Subsequently, mobility is affected by some aspects due to fuel utilization and emission of pollution. The data on release is substantial. Rakha and Ding [52] examine the outcome of vehicle stops on fuel utilization and emissions. While an implementation [52] of vehicle stop occurs, the vehicle fuel utilization and emission rate rise appropriately as per the research. Additionally, the vehicle stop intensity had considerable effects on vehicle emission rates at excessive crushing speed. The fuel utilization is more or less 80% when the empirical dataset is used to examine the effects of traffic congestion based on energy and outflow. Also, fuel utilization is lesser than on-trip time discovered by Terrier *et al.* [53]. The environmental performance is improved by focusing on crucial urban mobility aspects such as vehicles, road networks, and drivers.

3.10 Toward a Sustainable Ecosystem of ITS

It is very hard to calculate the essence of an intelligent transport system. It is tough to calculate the essence of an intelligent transport system in enhancing daily safety, mobility, and productivity. Data and communication sectors, travel systems and communication environments, integrated vehicles, and new trends such as IoT can fall under the shadow of intelligent transport technology. After the deep research in transport, communication, DBMS, and management, there is still a largely unsolved issue that impedes a wide range of use of advanced ITS. New research presented that blockchain is a growing technology that permits decentralized coordination. It can defeat intrinsic ITS aspects such as security and scalability. However, none of

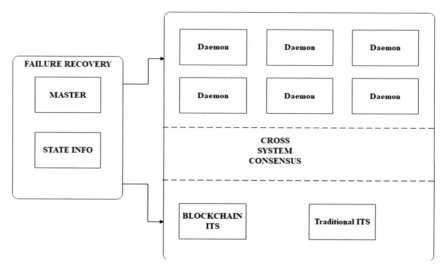

Figure 3.6 ITS ecosystem architecture.

these researches highlighted the tricky question: How can we develop a long-term ITS ecosystem? This section shows an outline of long-term ITS ecosystem.

The overall representation of ITS ecosystem architecture [54] is shown in Figure 3.6. It is constructed on a foundation of blockchain-oriented and traditional ITS systems and a common system consensus (CSC) abstraction and daemons on top. The level of daemons is raised since the compatibility barrier is solved. The lost recovery procedure is handled by the component present on the left. It needs direct contact with daemons and the ITS system only. Common system consensus protocols are atomic by an outline, which means that they either change the state of all ITS or the state remains constant. It is not needed to regain from loss as per the results. This architecture is a core environment that is available to the public. Compared to TCP/IP or OSI reference model, it allows any protocol to succeed the functionality and indicates the interactions between the components. Figure 3.6 shows some of the specifications of every component design.

3.10.1 Common system consensus and decentralization

A large number of ITS systems must be integrated and joined to complete the decentralization. Like the TCP/IP model, the protocol assures that end-to-end data transmission is the only way. While working with the ITS sector, a more robust abstraction is needed due to the intricacy of interaction. Let us consider a number of blockchain-dependent ITS systems and its coin or token. Due to some coin demerits such as time consumption to get, exchange, and work, this state may show a barrier for users who want to integrate with all of them. This results in simply offering delivery that will make enhancing software prohibitively critical. Thankfully, the fundamental aspects of coordination across several common system consensus protocols must offer the following end-to-end assurance in general:

a. The condition is accomplished if all users follow the protocol truthfully
b. No conforming user is harmed if a few users deviate from the protocol
c. No coalition has an incentive to deviate from the protocol

3.10.2 Daemons and backward compatibility

A daemon is a background running process, offering a few primary services to another process. It operates away from expired vehicles, preparing the ecosystem to be backwards compatible. The principal preparation is to have many daemon processes working on the many high-level automobiles. Each will interface with the ITS sector as a part of a heterogeneous old automobile. Specifically, every old automobile must register for the daemon service and purchase some credits in advance via web service. On another side, the advanced vehicles will first utilize computer vision and vehicle-to-vehicle contact to find the neighboring traditional vehicles. Next, those advanced vehicles can agree upon the responsibility of the daemon process. If they are not enough, the advanced aging vehicles may be unable to interact with the ITS system. It has two ramifications: i) ITS allows entry into an erroneous state due to unexpected actions; ii) ITS uses an auxiliary method to process those cars. Although it is an improvement, it is losing efficiency. So, for this case, we go for the third component.

3.10.3 Failure recovery and self-stabilization

They keep tracking the current state, and rolling back to most of the new proper states is essential to achieve a self-stabilizing ecosystem. Within the blockchain-based system, we store the pertinent state data, including states from both daemons and ITS systems. The examined methodology is performed regularly by the controller process. If it detects an incorrect state, it takes the following actions: i) scan the previous state to find a snapshot, i.e., consistent with all related ITS system and daemons; ii) notify them to roll back to the previous state snapshot; iii) identify affected vehicle and work out on recovery procedures like sending bills, refunds, etc.; iv) collect garbage. It is to remove unnecessary state data. A long-lived smart contract on the blockchain is used to implement the primary process.

3.10.4 Importance of sustainable development goal in transportation sectors addressing intelligent system

With its Sustainable Development Goals, the 2030 Agenda for Sustainable Development has been considered interconnected and indivisible since its beginning. This knowledge has led to review committees of the essence and resilience of interlinkages throughout goals and targets to identify areas of strategic significance for application and show how initiatives can conceivably magnify effects across different 2030 agendas, such as the climate goals. However, as recently underscored by the 2019 Global Sustainable Development Report (GSDR), rapid progress in these areas is only achievable if the systems connecting the SDG objectives and targets are altered in a manner that can solve trade-offs and deliver on potential synergies. The GSDR identified six entry points where action is required to achieve systemic change and sustainable development. Several such well-being and abilities, energy, cities, and food systems are intertwined with sustainable transportation, highlighting its promise as a cross-sectoral accelerator of the SDGs and effective climate action. Some SDGs, such as SDG target 3.6 on road safety, SDG target 9.1 on the power grid, and SDG target 11.2 on supplying access to safe, affordably priced, accessible, and sustainable transportation systems for all and expanding public transportation, are directly linked to sustainable mobility through specific targets and indicators. Several other SDGs were indirectly linked to sustainable transportation because of their enabling function. Sustainable and resilient infrastructure, for example, is expected to influence the achievement of up to 92% of all SDG targets, including but not limited to transportation. Likewise, attaining additional Sustainable Development Goals (SDGs) and associated objectives can contribute to the development and enhancement of sustainable transportation systems by raising household incomes, enhancing capacity for utilizing data, enabling greater accessibility, and implementing innovative technological approaches aimed at reducing the impact of climate and other environmental factors. To accelerate progress towards achieving Sustainable Development Goal (SDG) targets and climate objectives, it is necessary for multiple stakeholders to carefully consider the interdependencies among nations and stakeholders as well as the interconnections between sustainable transportation and various SDGs and targets during the planning, asset management, and implementation stages [55].

3.10.5 Renewable/green energy impact in intelligent transport and smart energy systems

A city may manage its resources more smartly with the Web of Things and different information and communications technology, forming the urban development concept of the future city [56]. It encourages citizen participation while also allowing for easier use of physical infrastructure. Smart energy and smart transport are two significant parts of the smart city, and electric vehicles (EVs) are expected to play a significant role. Energy supplies or the power grids power electric vehicles. An extensive fleet of EVs may be charged via charging points and parking facilities with correct scheduling. Although a single EV's battery capacity is limited, a group of EVs can act as an ample power supply or load, forming a vehicle-to-grid (V2G) arrangement. EVs can support the grid in V2G by providing various real-time responses and auxiliary services and obtaining energy from the grid [57]. We can minimize our reliance on fossil fuels and better use renewable energy. The electric vehicle market is rapidly expanding, and many EVs will undoubtedly be on the road shortly. EVs are also crucial components in the development of intelligent transportation systems. Self-driving cars (AVs) and the systematic remote control of AV fleets will expand the scope of smart energy and intelligent transportation systems [58].

3.11 Conclusion

According to the findings, ITS applications can improve the relative efficiency of urban traffic vs. economy, energy, and the environment, reduce total road transport energy usage, generate a cost of time savings, reduce journey time, reduce fuel consumption, reduce non-renewable energy use, as well as reduce traffic congestion This demonstrates that ITS can enhance the energy efficiency of a building. System of transportation. ITS is a long-term, ever-improving implementation in a country to deliver low-cost, people-centered transportation and environmentally friendly public transit systems. As a result, ITS should be prioritized and promoted because it is one of the essential factors in improving its energy efficiency. As new technologies emerge, ITS applications can be modified to maximize energy efficiency. ITS surveillance, communication, and information are all improving over time.

References

[1] S. C. Davis, S. W. Diegel, and R. G. Boundy, *Transportation Energy Data Book: Edition 30*, 30th ed. Oak Ridge, Tennessee: Oak Ridge National Laboratory, 2011.

[2] J. He, Z. Zeng, and Z. Li, "An Analysis on Effectiveness of Transportation Demand Management in Beijing," *Journal of Transportation Systems Engineering and Information Technology*, vol. 9, no. 6, pp. 114–119, Dec. 2009.

[3] J. Sun, Q. Liu, and Z. Peng, "Research and Analysis on Causality and Spatial-Temporal Evolution of Urban TrafficCongestions—A Case Study on Shenzhen of China," *Journal of Transportation Systems Engineering and Information Technology*, vol. 11, no. 5, pp. 86–93, Oct. 2011.

[4] J. He, Z. Zeng, and Z. Li, "Benefit Evaluation Framework of Intelligent Transportation Systems," *Journal ofTransportation Systems Engineering and Information Technology*, vol. 10, no. 1, pp. 81–87, Feb. 2010.

[5] *WHITE PAPER (European Transport Policy for 2010_: Time to Decide)*. Office for Official Publications of theEuropean Communities, 2001

[6] MIROS. (2011). General Road Accident Data in Malaysia (1995 – 2010). Retrieved 9 August, 2012, from http://www.miros.gov.my/web/guest/road

[7] N. Shah, S. Kumar, F. Bastani, and I.-L. Yen, "Optimization Models for Assessing the Peak Capacity Utilization ofIntelligent Transportation Systems," *European Journal of Operational Research*, vol. 216, no. 1, pp. 239–251, Jan.2012.

[8] K. He, H. Huo, Q. Zhang, D. He, F. An, M. Wang, and M. P. Walsh, "Oil Consumption and CO2 Emissions in China'sRoad Transport: Current Status, Future Trends, and Policy Implications," *Energy Policy*, vol. 33, no. 12, pp. 1499–1507,Aug. 2005.

[9] S. Jia, H. Peng, S. Liu, and X. Zhang, "Review of Transportation and Energy Consumption Related Research," *Journalof Transportation Systems Engineering and Information Technology*, vol. 9, no. 3, pp. 6–16, Jun. 2009.

[10] C. Ming and S. Yuming, "Agent Based Intelligent Transportation Management System," *2006 6th InternationalConference on ITS Telecommunications Proceedings*, pp. 190–193, Jun. 2006.

[11] Q. Shi and W. Shen, "Development of Integrated Information Platform for Intelligent Transportation Systems," *2003IEEEI Proceeding on Intelligent Transportation Systems*, vol. 1, pp. 54–59, 2003.

[12] L. Tsu-tian, "Research on Intelligent Transportation Systems in Taiwan," *Proceedings of the 27th Chinese ControlConference*, pp. 18–23, 2008.

[13] C. R. Berger and E. Smith, "Intelligent Transportation Systems Provide Operational Benefits for New YorkMetropolitan Area Roadways: A Systems Engineering Approach," *Systems, Applications and Technology Conference,2007. LISAT 2007. IEEE Long Island*, pp. 1–8, 2007.

[14] P. Papadimitratos, A. La Fortelle, K. Evenssen, R. Brignolo, and S. Cosenza, "Vehicular Communication Systems:Enabling Technologies, Applications, and Future Outlook on Intelligent Transportation," *IEEE CommunicationsMagazine*, pp. 84–95, 2009.

[15] Z. Wei, P. Zhao, and S. Ai, "Efficiency Evaluation of Beijing Intelligent Traffic Management System Based on Super-DEA," *Journal of Transportation Systems Engineering and Information Technology*, vol. 12, no. 3, pp. 19–23, Jun.2012.

[16] J. C. Miles, K. Chen. (ed.), Intelligent Transport Systems Handbook Publisher, Andrew Berrball, 2004.

[17] A. Shah and L. Jongdal, "Intelligent Transportation Systems in Transitional and Developing Countries," *IEEE A&ESystems Magazine on Aerospace and Electronic Systems*, pp. 27–33, 2007.

[18] A. Garcia-Ortiz, S. M. Amin, and J. R. Wootton, "Intelligent Transportation Systems- Enabling Technologies," *Mathematical Computer Modeling*, vol. 22, no. 4, pp. 11–81, 1995.

[19] J. R. Wootton, A. García-Ortiz, and S. M. Amin, "Intelligent Transportation Systems: A Global Perspective,"*Mathematical and Computer Modelling*, vol. 22, no. 4–7, pp. 259–268, Aug. 1995.

[20] Strategic Plan for Intelligent Vehicle-Highway Systems in the United States. IVHS America, 1992.*Energy Efficient Approach Through Intelligent Transportation System: A Review* 169

[21] Y. Sugawara, "Understanding the Differences in the Development and Use of Advanced Traveler Information Systemsfor Vehicles (ATIS/V) in the U.S., Germany, and Japan," Massachusetts Institute of Technology, 2007.

[22] ITS America. (2011). World Congress. Retrieved 9 August, 2012, from http://www.itsworldcongress.org/worldcongress.html

[23] https://its-uk.org.uk/wp-content/uploads/2017/02/ITS-UK-Benefits-of -ITS.pdf

[24] M. C. Coelho and N. Rouphail, "Assessing the impact of V2V/V2I communication systems on traffic congestion and emissions," in *Proceedings of the European Conference on Human Centred Design for Intelligent Transport Systems*, Berlin, Germany, April 2010.

[25] N. Haworth and M. Symmons, "Driving to reduce fuel consumption and improve road safety," Monash UniversityAccident Research Centre, 2001, http://acrs.org.au/files/arsrpe/RS010036.pdf.

[26] Nasir, Mostofa Kamal, et al. "Reduction of fuel consumption and exhaust pollutant using intelligent transport systems." *The Scientific World Journal* 2014 (2014).

[27] T. Nguyen, O. Berder, and O. Sentieys, "Cooperative MIMO schemes optimal selection for wireless sensor networks," IEEE 65th Vehicular Technology Conference, VTC-Spring 07, pp. 85–89, 2007.

[28] Nguyen, Tuan-Duc, et al. "Energy efficient cooperative communication techniques for Intelligent Transport System." *The 2011 International Conference on Advanced Technologies for Communications (ATC 2011)*. IEEE, 2011.

[29] https://www.transportpolicy.net/standard/singapore-feebate/

[30] Makarova, Irina, et al. "Development of sustainable transport in smart cities." 2017 IEEE 3rd International Forum on Research and Technologies for Society and Industry (RTSI). IEEE, 2017

[31] T. Mahlia, S. Tohno, and T. Tezuka, "International experience on incentive programing support of fuel economy standards and labelling formotor vehicle: a comprehensive review," Renewable and Sustainable Energy Reviews, vol. 25, pp. 18-33,2013.

[32] M. N. Uddin, W. M. A. W. Daud, H. F. Abbas, M. T. Islam, Z. Z. Chowdhury, and S. Das, "Effects of pyrolysis parameters on hydrogenformations from biomass," RSC Advances, vol. 4, no. 21, pp. 10467-10490, 2014.

[33] Automobile Emissions: An Overview, U.S. ENVIRONMENTAL PROTECTION AGENCY EPA 400-F-92-007 OFFICE OF MOBILE-SOURCES

[34] Imtiyaz, Shaikh HaqueMobassir, and Shaikh AbdurRehman Mohammed Sadique. "Intelligent transport systems a comprehensive way to regulate and curb vehicular pollution." *2015 International Conference on Communications and Signal Processing (ICCSP)*. IEEE, 2015.

[35] Otsu N.,"A thresholding selection using gray level histograms", IEEE /IEEE Transactions Systems/Man Cybernetics 1979, vol. 9, pp. 62-69

[36] Brendan T. Morris, M. M. Trivedi. "Learning, Modeling and Classification of Vehicle Track Patterns from Live Video". IEEETransaction on Intelligent Transportation Systems, vol. 9, issue 3, 425- 437, Sept. 2008.

[37] Rafael C. Gonzalez, R. E. Woods, S. L. Eddins. "Digital Image Processing Using Matlab". Pearson Prentice Hall, 2004.

[38] Harilakshmi, V. S., and P. Arockia Jansi Rani. "Intelligent vehicle counter-a road to sustainable development and pollution prevention (P2)." 2016 International Conference on Energy Efficient Technologies for Sustainability (ICEETS). IEEE, 2016.

[39] Cooperative mobility systems and services for energy efficiency, see ht tp://www.ecomove-project.eu/. 2015.

[40] ECOSTAND: Joint EU-Japan-US task force on the development of a standard methodology for determining the impacts of ITS on energy efficiency and CO2 emissions, see http://www.ecostandproject.eu/. 2015.

[41] EU Compass4D research project, see http://www.compass4d.eu/. 2015.

[42] Co-operative systems for sustainable mobility and energy efficiency, see http://www.cosmo-project.eu/. 2015.

[43] Co-operative networked concept for emission responsive traffic operations, see http://www.traffictechnologytoday.com/news.php?NewsID= 27116. 2014.

[44] Assessment Methodologies for ICT in multimodal transport from user behaviour to CO2 reduction, see http://www.amitran.eu/. 2015.

[45] Cooperative advanced driver assistance system for green cars, see http: //www.transport-research.info/web/projects/project_details. cfm?id=44395. 2015.

[46] Barth, Matthew J., Guoyuan Wu, and Kanok Boriboonsomsin. "Intelligent transportation systems and greenhouse gas reductions." *Current Sustainable/Renewable Energy Reports* 2.3 (2015): 90-97.

[47] Panday, Aishwarya, and Hari Om Bansal. "Green transportation: need, technology and challenges." International Journal of Global Energy Issues 37.5-6 (2014): 304-318.

[48] E. Ericsson, "Independent driving pattern factors and their influence on fuel-use and exhaust emission factors," *Transp. Res. D, Transp. Environ.*,1vol. 6, no. 5, pp. 325–345, Sep. 2001.

[49] M. van der Voort, M. S. Dougherty, and M. van Maarseveen, "A prototype fuel-efficiency support tool," *Transp. Res. C, Emerging Technol.*, vol. 9, no. 4, pp. 279–296, Aug. 2001.

[50] K. Brundell-Freij and E. Ericsson, "Influence of street characteristics, driver category and car performance on urban driving patterns," *Transp. Res. D, Transp. Environ.*, vol. 10, no. 3, pp. 213–229, May 2005.

[51] K. Boriboonsomsin and M. Barth, "Impacts of road grade on fuel consumption and carbon dioxide emissions evidenced by use of advanced navigation systems," *Transp. Res. Rec., J. Transp. Res. Board*, vol. 2139, no. 1, pp. 21–30, 2009.

[52] H. Rakha and Y. Ding, "Impact of stops on vehicle fuel consumption and emissions," *J. Transp. Eng.*, vol. 129, no. 1, pp. 23–32, Jan./Feb. 2003.

[53] M. Treiber, A. Kesting, and C. Thiemann, "How much does traffic congestion increase fuel consumption and emissions? Applying fuel consumption model to NGSIM trajectory data," presented at the Transportation Research Board 87th Annual Meeting, Washington, DC, USA, 2008, Paper 08-2715

[54] Tseng, Lewis, and Liwen Wong. "Towards a sustainable ecosystem of intelligent transportation systems." 2019 IEEE international conference on pervasive computing and communications workshops (PerCom Workshops). IEEE, 2019.

[55] Maria E. Mondejar, Digitalization to achieve sustainable development goals: Steps towards a Smart Green Planet, Science of The Total Environment, Volume 794, 10 November 2021, 148539

[56] Xiaoyi, Zhang, Wang Dongling, Zhang Yuming, Karthik Bala Manokaran, and A. Benny Antony. "IoT driven framework based efficient green energy management in smart cities using multi-objective distributed dispatching algorithm." Environmental Impact Assessment Review 88 (2021): 106567.

[57] Zhu, Lingling, Jie Fangi Shi, Yi Hai Shi, Hai Peng Xu, A. Shanthini, and Tamizharasi G. Seetharam. "Renewable Green Energy Resources for Next-Generation Smart Cities using Big Data Analytics." Journal of Interconnection Networks (2021): 2141004.

[58] Pandian, Shunmugham R. "Intelligent Mechatronic Technologies for Green Energy Systems." In 2010 GSW. 2022.

4

Application of ANN Techniques to Mitigation of Power Quality Problems

Alka Singh

Department of Electrical Engineering, DTU, India
E-mail: alkasingh@dce.ac.in

Abstract

Significant research in power system techniques and controls has revolutionized the world. Now, it is time to integrate artificial intelligence (AI) techniques into the power sector for even more intelligent and efficient systems. In the last few decades, power engineers have effectively designed intelligent controllers to mitigate power quality problems and challenges. Application of power electronics to power systems, renewable energy integration, and design of high-voltage DC transmission systems, flexible AC transmission systems, and custom power devices has been a remarkable journey.

The new norm is power quality (PQ) improvement and mitigation of problems using AI techniques. The search for fast, cost-effective solutions rests on new learning techniques for solving perennial PQ problems. It includes designing neural networks (NN), fuzzy-based solutions, and hybrid combinations such as ANFIS. The literature review suggests several new and effective control techniques designed to mitigate PQ problems.

This chapter starts with some conventional and new AI-based solutions for PQ improvement. Several newly developed functional NN techniques are discussed briefly, such as basic FLANN, trigonometric FLANN, Legendre's NN, and time-delayed recurrent NN. After that, the application of the designed NN controllers in reactive power compensation and PQ problem mitigation is discussed. The PQ problems discussed in the chapter

73

include power factor (pf) correction, load balancing, and reactive power (Q) compensation. Two case studies are included – (i) without PV integration and (ii) with PV integration. Precise control and MATLAB models are discussed, and results are presented for a single/three-phase power distribution system encountering several PQ problems. Further, the chapter discusses new NN techniques that can be explored for PQ mitigation in future work.

Keywords: Compensation, power quality, photovoltaic array, power factor correction, renewable energy.

4.1 Introduction

Power quality is a relevant topic discussed among power engineers. Due to the ever-increasing PQ problems, engineers and scientists are working on practical solutions. Both single-phase and three-phase distribution systems face poor voltage regulation, poor power factor (pf), harmonics, load unbalancing, high neutral currents, and the presence of sags, swells, and grid distortion [1]. A single and effective solution to all the above-mentioned problems is not feasible. Therefore, several solutions have been suggested in the last decade, viz. series, shunt, and hybrid connection of compensators. A reasonable solution needs to be worked out depending on the incidence of problems. A shunt compensator is designed and installed to mitigate load compensation, while problems in the utility grid such as voltage sag, swell, and distortion are mitigated using series devices. A unified operation is the most effective solution to all PQ problems but also the most costly. The age-old conventional solutions in the form of fixed filters also effectively provide the required reactive power compensation. However, these get detuned and less effective over a while.

This chapter discusses inverter-based custom power devices, which have proved to be highly effective and helpful in mitigating PQ problems at the load end. The shunt-connected device can be suitably controlled to solve PQ problems. This shunt compensator can have many configurations and be designed using different power electronic switches such as insulated gate bipolar transistors (IGBTs), thyristors, and other switches. Practically the H-bridge configuration is realized for a single-phase distribution system, while a three-leg configuration with six IGBT devices and antiparallel diodes is commonly employed. These active filters can be effectively controlled to provide reactive power compensation and PQ mitigation. The active filter is controlled to inject a calculated magnitude of the current to nullify the harmonics present in the load current. The following PQ problems can be effectively solved using custom power devices:

- reduction of harmonics in the grid;
- correction of power factor to almost unity;
- providing load balancing in a three-phase system;
- providing reactive power support.

The role of control techniques in solving these PQ problems is immense. Research carried out in the last decade highlights the role of effective control algorithms. New and more effective controllers and adaptive algorithms have replaced conventional ones [1]. Several authors have discussed fuzzy and NN-based controllers for active filters and work on combining these two as an ANFIS controller. This chapter aims to focus on a particular class of NN techniques that are not commonly described in the literature.

Moreover, the NN techniques are adaptive so that weights get converged online automatically. This is done so that the self-adaptive process can be easily implemented on the hardware using low-cost digital signal processors. Thus, the control of the inverters using NN techniques is feasible and effective.

The backpropagation (BP) NN technique is a popular and well-known technique based on updating weights from one layer to another [2]. However, there are problems with this complex method, and, hence, new NN techniques must be devised. Multiple-layer NN is also feasible and commonly employed to train complex problems. However, single-layer-based NN structures are also efficient and can be thoroughly trained to produce desired results. Further, just as the choice and number of neurons in the single layer can be decided as per the complexity of the problem, these NNs can be designed using many expansion terms. The use of functional expansion of inputs using trigonometric terms, algebraic terms, polynomial functions, etc., is possible and is the focus of this work.

A category of functional layer NN techniques (FLANN) is employed for varied applications, from stock prediction to load forecasting [3, 4] The backpropagation technique for mitigating PQ problems is presented in [5]. FLANN represents a single-layer NN whose expansion can be based on trigonometric [6], Chebyshev [7], or other polynomials like Legendre [8]. All functional layer networks utilize a similar structure and are based on the non-linear functional expansion of the input vector. A time-delay-based recurrent neural network (RNN) technique [9] is also possible. The functional expansion needs to be trained to capture the weights or the non-linear relationship between the inputs and outputs. A multiple-layer-based NN performs the same task but uses several hidden layers in between the input and the output layers. Hence, the use of single-layered NN is preferred. The least mean square (LMS) technique is a convenient method for updating the network weights and biases [10]. Some recent papers on the application of NN for load forecasting [11] and ANN predictive control of power converters [12] are mentioned in the literature.

This chapter focuses on the design and development, implementation, and performance analysis of several functional NNs used for mitigating PQ problems. The emphasis of the chapter is on trigonometric FLANN (T-FLANN), Legendre polynomial-based FLANN (L-FLANN), and recurrent NN (RNN). The main objectives of this chapter include:

- highlighting PQ problems in single/three-phase power distribution systems;
- designing shunt controller for mitigation of PQ problems;
- designing effective and self-adaptive NN techniques for effective PQ control;
- focusing on T-FLANN, L-FLANN, and RNN networks as effective solutions;
- extension of the developed techniques for control of PV integrated system.

4.2 ANN Techniques and Analysis

The functional artificial neural network (FLANN) is based on a single configuration comprising a single layer apart from the standard input and output layers. This layer contains some practical expansion terms of the input signal shown in Figure 4.1. The motive of using functional expansion is to increase the boundaries of the input vector to learn and understand the network in a better and more comprehensive manner. This also provides the NN with higher capability and efficient operation. It further overcomes the challenges in selecting the required number of hidden layers as in a multi-layered NN. The simple network can be trained easily using the backpropagation technique or the gradient descent approach. The simplified structure of the NN is also desirable in decreasing computational time and complexity. The algorithm development, implementation, and hardware testing become feasible.

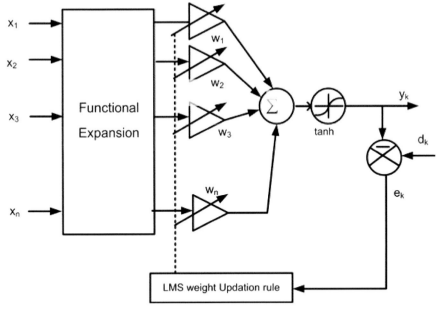

Figure 4.1 Typical representation of FLANN.

Different FLANN techniques are mentioned in the literature, but a detailed investigation and application of these in power distribution systems are discussed in this section.

4.2.1 Mathematical formulation of generalized FLANN

Figure 4.1 shows the basic structure of the adaptive FLANN technique. Based on the expansion terms chosen for a particular network, different types of FLANNs can be designed. The self-organization and real-time updation of weight are significant attractions of these techniques. The training of NN is of paramount significance, and to achieve this, several training algorithms may be selected, viz. backpropagation or variants of the least mean square (LMS) algorithm based on the gradient descent approach [10]. These involve the repeated updating of weights until convergence is achieved. The error e_j between the desired (y_{actual}) and estimated quantity (y_{est}) needs to be minimized, as shown by the following equation:

$$e_j = y_{\text{actual}} - y_{\text{est}}. \tag{4.1}$$

Several variants of the LMS algorithm are available in the literature to update the weights (w). However, a simplified weight updation technique is quite effective and preferred, as given below:

$$w(k+1) = w(k) + \mu e(k)\sin\theta, \tag{4.2}$$

where$^{\mu}$ denotes the acceleration factor and $\sin\theta$ is the unit in phase template.

4.2.2 Mathematical formulation of trigonometric FLANN (T-FLANN)

Different functional expansion models in the form of power series expansion, expansion involving trigonometric terms, Chebyshev polynomial expansion based, or Hammerstein-based functional expansion can be implemented quickly. The functional expansion involving only trigonometric terms results in the design of a T-FLANN structure as in Figure 4.2. Its weight vectors can be updated and optimally computed for updating the fundamental component of load current. The learning of this functional T-FLANN [6] is described in the following equation:

$$i_{Lest} = \Sigma_{i=1}^{M} w^{T} s_{i}, \tag{4.3}$$

where i_{Lest} is the fundamental component of non-linear load current corresponding to 50 Hz. Non-linear functional trigonometric expansion involves non-linear trigonometric functional expansion s_i described in eqn (4.3)

$$i_L = [i_L(n), i_L(n-1), \ldots, i_L(n-N+1)]^{T}, \tag{4.4}$$

$$s_i = f(i_L(k)), \quad for \ 1 \leq i \leq M, \ 1 \leq k \leq N, \tag{4.5}$$

where $w = [w_1, w_2, w_3, w_4, \ldots, w_M]^{T}$ is a set of weights corresponding to each term. The expansion of the T-FLANN based on trigonometric expansion terms is indicated below. The input signal is the load current $i_L(n)$

$$s_{1,1} = i_L(n), s_{1,2} = \sin[\pi i_L], s_{1,3} = \cos[\pi i_L], s_{2,1} = \sin[2\pi i_L],$$

$$s_{2,2} = \cos[2\pi i_L].$$

Similarly, $s_{N,M-1} = \sin[h i_L]$ and $s_{N,M} = \cos[h\pi i_L]$ correspond to terms for "h" order of harmonics. Further, the effect of cross-terms can also be considered in the design of T-FLANN. However, a simplified expression with only a few dominant terms is sufficient and preferred in this work.

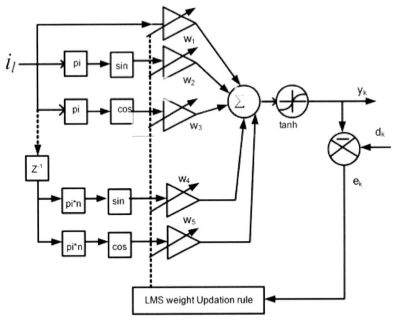

Figure 4.2   Typical representation of T-FLANN.

4.2.3 Mathematical formulation of Legendre-FLANN (L-FLANN)

The L-FLANN algorithm is based on the standard Legendre polynomials [8] that are the solutions of the Legendre differential equation given below:

$$\left(1 - x^2\right) \frac{d^2 y}{dx^2} - x \frac{dy}{dx} + n\left(n + 1\right) y = 0 \tag{4.6}$$

where "n" denotes the degree of the polynomial, and y is the solution of the differential equation in terms of variable x. The series converges for $|x| < 1$, i.e., in the interval for $-1 < x < 1$, for all values of n.

The series solution is given in the form of

$$y = \Sigma_{k=0}^{\infty} b_k x^k , \tag{4.7}$$

where b_k denotes the general coefficient terms.

Differentiating the equation and solving gives the coefficients recursively

$$b_{k+2} = \frac{k\left(k + 1\right) - n(n + 1)}{\left(k + 2\right)\left(k + 1\right)} b_k. \tag{4.8}$$

All the coefficients are recursively computed in terms of b_0 and b_1, which are taken as arbitrary constants. Substituting $k = 0, 1, 2, 3, \ldots$ gives

$$b_2 = \frac{-n(n+1)}{2!} b_0, \quad b_3 = \frac{-(n-1)(n+2)}{3!} b_1, \text{ etc.}$$

Inserting the value of coefficient in eqn (4.8), we get

$$y = b_0 \left[1 - \frac{n(n+1)}{2!} x^2 + \frac{(n-2)(n)(n+1)(n+3)}{4!} x^4 - \cdots \right]$$
$$+ b_1 \left[x - \frac{(n-1)(n+2)}{3!} x^3 \right.$$
$$+ \frac{(n-3)(n-1)(n+2)(n+4)}{5!} x^5 - \cdots \left. \right], \tag{4.9}$$

$$y = b_0 y_1 + b_1 y_2, \tag{4.10}$$

where the series y_1 contains even powers of x, while the series y_2 contains odd powers of x only and y_1 and y_2 are linearly independent of solutions of eqn (4.8). For a particular choice of b_0 and b_1, Legendre polynomial of order n results, which is denoted by $p_n(x)$.

Here, b_0 and b_1 have been selected to be

$$b_0 = \frac{(-1)^{n/2} 1.3.5 \ldots (n-1)}{2.4.6 \ldots n}, b_1 = \frac{(-1)^{n-1/2} 1.3.5 \ldots n}{2.4.6 \ldots n - 1}. \tag{4.11}$$

Thus, we get

$$p_n(x) = \frac{1}{2^n n!} \frac{d^n}{dx^n} (x^2 - 1)^n. \tag{4.12}$$

The first few Legendre polynomials (p_n) commonly used are

$$p_1(x) = 1, \; p_2(x) = x, \; p_3(x) = \frac{1}{2}[1 - 3x^2], \; p_4(x) = \left(\frac{1}{2} \right) [5x^3 - 3x]$$

where the input is "x." Now, if the input is considered as the load current i_L, Legendre polynomial terms are depicted as

$$i_{L0}(x) = 1, \; i_{L1}(x) = i_L, i_{L2}(x) = \frac{1}{2}[1 - 3i_L^2], \; i_{L3}(x) = \left(\frac{1}{2} \right) [5i_L^2 - 3i_L].$$

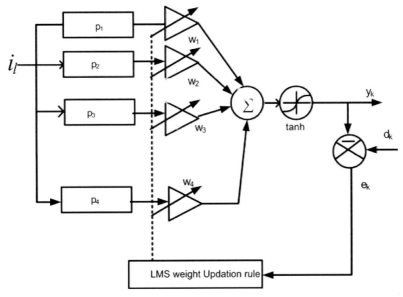

Figure 4.3 Typical representation of L-FLANN.

An appropriate number of terms (three or four) can be considered to efficiently depict the dominant terms and train the network further for obtaining the load component corresponding to 50 Hz. A representation of L-FLANN is depicted in Figure 4.3.

4.2.4 Mathematical formulation of recurrent NN (RNN)

The structure of time-delayed recurrent NN is discussed next, and it is essentially focused on extracting the fundamental weight component of the load current signal. Its structure comprises an input layer, an output layer, and a hidden layer. An adaptive version of this algorithm for updating the weights is presented, which can be based on the backpropagation technique or any variant of the LMS technique. A simplified LMS-based real-time learning technique is also selected here, which is quite effective.

The structure of the RNN is shown in Figure 4.4, which also shows a sigmoid activation function. The output of the time-delayed RNN algorithm [9] is expressed as

$$y_k = f\left(w_1\left(x_k w_3 + w_l\left[\begin{array}{ccccc} y_{k-1} & y_{k-2} & y_{k-3} & \cdots & y_{k-n} \end{array}\right]\right)\right) \times w_2,$$
$$(4.13)$$

where y_k is the output signal, y_{k-n} represents "*n*" time-delayed signals of output, and x_k is the input signal. Moreover, w_1, w_2, and w_3 are selected to be fixed weights and chosen as constants. One of the weights, w_l, is updated recursively, which is done using the LMS algorithm. It is shown for the estimation of weight w_l as

$$w_l(k+1) = w_l(k) + \alpha e_k x_k. \tag{4.14}$$

The RNN is designed for the estimation of the fundamental load component of the non-linear load current for PQ mitigation. This estimation of the fundamental current component uses an error (e_k) term expressed as

$$e_k = i_L - i_{L,\text{est}}. \tag{4.15}$$

The fundamental load current component (I_{kl}) and updated weight (w_1) are expressed as

$$y_k = f\left(w_1\left(u_p w_3 + w_1\begin{bmatrix} y_{k-1,l} & y_{k-2,l} & y_{k-3,l} & \cdots & y_{k-n,l} \end{bmatrix}\right)\right) \times w_2. \tag{4.16}$$

The estimated fundamental weight (y_k) is used for the generation of reference supply currents and the gating pulses, and α represents the acceleration factor used for weight updation.

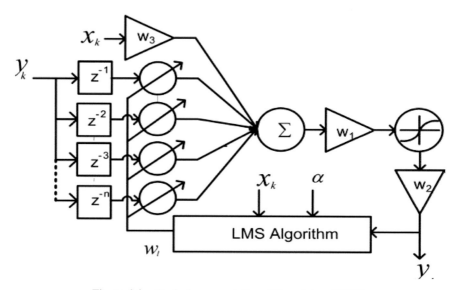

Figure 4.4 Typical representation of time-delayed RNN.

4.2.5 Mathematical formulation for inverter switching loss computation

The design of an effective controller for closed-loop operation needs to include the computation of the switching losses. These losses need to be estimated and met from the grid. The error between the sensed and reference DC bus voltage of the shunt compensator is computed as e_{dc} and provided to the PI controller to meet the switching losses (w_{loss}) of the compensator and maintain the DC-link voltage to the reference value

$$e_{\text{dc}} = v_{\text{dc}}^* - v_{\text{dc}}(n) \tag{4.17}$$

$$w_{\text{loss}}(k) = w_{\text{loss}}(k-1) + K_p \{e_{\text{dc}}(n) - e_{\text{dc}}(n-1)\} + k_i e_{\text{dc}}(n) \tag{4.18}$$

where K_p and k_i are the proportional and integral gain of the PI controller.

The total active power requirement of the load is met from the grid and it is computed by adding the weighted fundamental load component of the current (w_{Lest}) and w_{loss} .

$$w_{\text{total}} = w_{\text{Lest}} + w_{\text{Loss}}. \tag{4.19}$$

The integration of PV to the DC link of the inverter further modifies eqn (4.19) to

$$w_{\text{total}} = w_{\text{Lest}} + w_{\text{Loss}} - w_{\text{pv}}. \tag{4.20}$$

Here, w_{pv} is called the PV feed-forward component term. It is subtracted from the sum of the active power load requirement and depends on the rating of the PV. Moreover, this feedback term is calculated differently for single-phase and three-phase PV interfaced systems.

4.3 Performance and Results

This section discusses the simulation results for some NN techniques discussed above. The objective of developing these techniques is to mitigate PQ problems such as pf correction, harmonic elimination, and providing reactive power ($+Q$) support. For, a three-phase system, the controller also helps to achieve load balancing on the grid side irrespective of the nature of the load. This is an additional functionality not realized in single-phase power distribution systems. Hence, a mix of single-phase and three-phase systems have been considered and designed with NN-based controllers. Further, the presence of PV support at the output of the DC link of the compensator provides an additional real power injection capability. Thus, both aspects, viz. without/with PV support, have been discussed in the results below.

4.3.1 Results with T-FLANN algorithm without/with PV integration

The designed TFLANN controller has been developed and simulated using the MATLAB-based Simulink model. A single-phase 110-V, 50-Hz supply feeds a non-linear load feeding a resistive–inductive branch at the DC terminals. The closed-loop shunt compensation results are first presented in Figures 4.5 and 4.6 without PV integration. The result shows the plots for supply voltage (V_s), load current (I_l), supply current (I_s), compensator current (I_c), and DC-link voltage (V_{dc}). From $t = 0.2$ s to $t = 0.4$ s, the non-linear load on the system is increased, which leads to a corresponding dip in the DC-link voltage V_{dc}. The PI controller on the DC link regulates the voltage to the set reference value of 200 V. The supply voltage and supply current are in-phase relationships indicating unity pf operation. This fact is also evident in Figure 4.6, which shows that the grid supplies no reactive power. The entire reactive power demand of the load is met wholly by the compensator. Since there is no additional source at the DC link to meet the active power demand of the load, the grid meets the entire active power requirement of the load. Before the load change at $t = 0.2$ s, the load demand is approximately 200 W, 70 vars. The control scheme is designed to control the shunt compensator to inject the entire Q demand of the load. Thus, as evident from Figure 4.6, almost no active power is injected by the compensator, and no reactive power

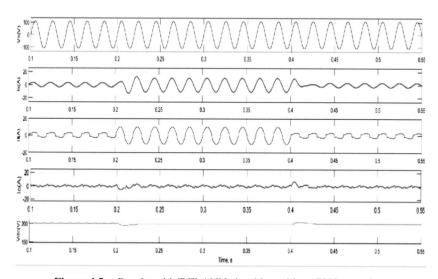

Figure 4.5   Results with T-FLANN algorithm without PV integration.

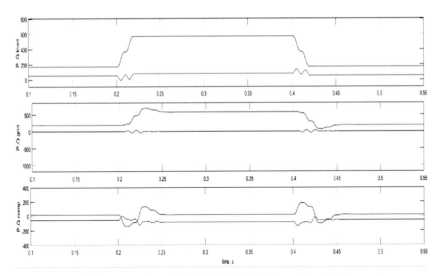

Figure 4.6 Power plots with T-FLANN algorithm without PV integration.

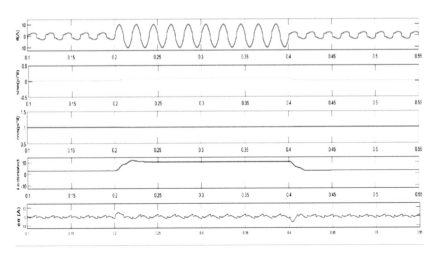

Figure 4.7 Results with T-FLANN algorithm showing intermediate signals.

demand is injected by the grid. This control objective remains the same even when load addition is performed from $t = 0.2$ to $t = 0.4$ s.

The intermediate results for the designed T-FLANN algorithm are first presented in Figure 4.7. The control algorithm can extract the fundamental current (I estimated) from the distorted load current. The result shows the load current (I_l) plots, dominant T-FLANN terms, viz. sine and cosine of load current. The error between the actual and estimated current is also observed, which happens to be relatively small. Moreover, as the load varies from $t =$ 0.2 to 0.4 s, the estimated fundamental current also increases proportionately. Thus, the algorithm works well, and the intermediate results demonstrate the satisfactory operation of T-FLANN in mitigating PQ problems.

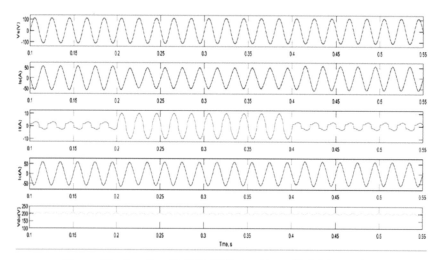

Figure 4.8   Results with T-FLANN algorithm with PV integration.

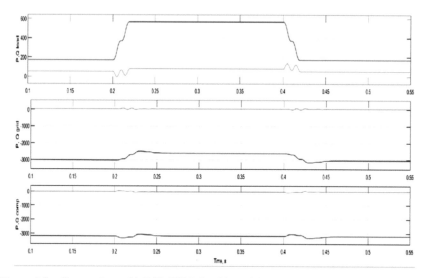

Figure 4.9   Power plots with T-FLANN algorithm with PV integrated single-phase system.

With PV integration, the closed-loop shunt compensation results are now presented in Figures 4.8 and 4.9. The PV source is integrated at the DC link of the H-bridge inverter. Figure 4.8 shows the plots for supply voltage (V_s), load current (I_l), supply current (I_s), compensator current (I_c), and DC-link voltage (V_{dc}). This figure shows the load current is smaller, having a peak of 10 A maximum, while the supply current and compensator current are pretty higher. This happens as the PV source injects maximum power into the grid since the load demand is low. Moreover, the supply voltage and current are not in the same phase. From $t = 0.2$ to $t = 0.4$ s, the non-linear load on the system is increased, leading to a corresponding dip in the DC-link voltage V_{dc} but stabilizes due to PI controller action. The PI controller on the DC link regulates the voltage to the set reference value of 200 V even during load variations. The power plots are reflected in Figure 4.9 for a single-phase grid-connected system. This fact is also evident in Figure 4.6, which shows that the grid supplies no reactive power. The load demand is 200 W and 70 vars and the PV capacity is approximately 3.2 kW. Thus, nearly 3 kW of active power is injected into the grid as the load is 0.2 kW. Here, the designed T-FLANN controller for the compensator provides reactive and active power support per the PV system's rating at the DC-link terminals.

4.3.2 Results with L-FLANN algorithm without/with PV integration

The designed L-FLANN controller has been developed next and simulated using the Matlab-based Simulink model. The system is the same as before and comprises a single-phase 110-V, 50-Hz supply feeding a non-linear load.

The closed-loop shunt compensation results with the Legendre-FLANN controller are first presented in Figures 4.10 and 4.11 without PV integration. The result shows the plots for supply voltage (V_s), load current (I_l), supply current (I_s), compensator current (I_c), and DC-link voltage (V_{dc}). From $t = 0.2$ to $t = 0.4$ s, the non-linear load on the system is increased, which leads to a corresponding dip in the DC-link voltage V_{dc}. The PI controller on the DC link regulates the voltage to the set reference value of 200 V. The supply voltage and supply current are in-phase relationships, indicating there is unity pf operation. This fact is also evident from Figure 4.11, which shows that the grid is supplying no reactive power.

The load demand is approximately 1250 W, 500 vars. At $t = 0.2$ s, the load demand increased to 2000 W and approximately 600 vars. After that, the same load is restored at $t = 0.4$ s. Figure 4.11 shows the three power plots

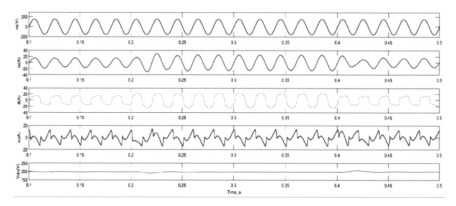

Figure 4.10   Results with L-FLANN algorithm without PV integration.

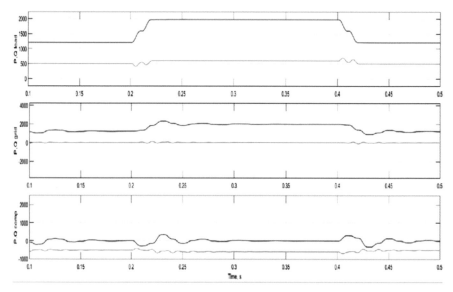

Figure 4.11   Power plots with L-FLANN algorithm without PV integration.

for the load, grid, and compensator. Both real power (red color) and reactive power (blue color) plots have been shown in Figure 4.11. It is observed that almost the entire reactive power demand of the load is met wholly by the compensator, and the grid is free from supplying any additional Q. Moreover, since there is no additional source at the DC link to meet the active power, the grid injects the active power demand of the load entirely.

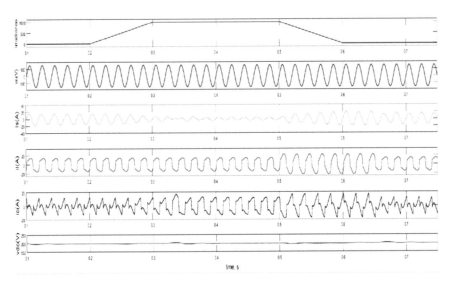

Figure 4.12 Results with L-FLANN algorithm with PV integrated system under irradiance and load change.

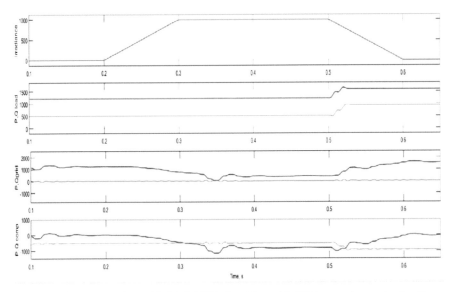

Figure 4.13 Power plots with L-FLANN algorithm with PV integration under irradiance variations.

Figures 4.12 and 4.13 show the results with the L-FLANN controller designed for the single-phase system integrated to the PV source. No DC−DC boost converter is placed between the PV system and the voltage source converter. Thus, the single-stage connection of PV is used. Moreover, the effects of irradiance change and load change are included in the results. A ramp increase of irradiance from 0 to 1000 W/m^2 is applied from 0.2 to 0.3 s followed by a ramp decrease of irradiance from 1000 W/m^2 to 0. Load is changed from the duration 0.5 to 0.65 s. Analysis of results in Figure 4.12 shows that as the irradiance level is ramped up, more current is injected by the compensator. The load current is constant till 0.5 s. The PV has a capacity of nearly 0.85 kW; when the irradiance is maximum, active power is injected the highest. The load has a higher demand of 1250 W; hence, the net active power injected from the grid is ∼400 W. Further, it is observed that the grid does not feed any additional reactive power (blue color plot) to the load. The control algorithm is designed to achieve this task of reactive power injection.

4.3.3 Results with RNN algorithm without/with PV integration

The designed Legendre-FLANN controller has been developed next and simulated using the Matlab-based Simulink model. The system considered is a three-phase 110-V ($L−L$), 50-Hz supply feeding a non-linear load.

The closed-loop shunt compensation results with the RNN controller are first presented in Figures 4.14 and 4.15 without PV integration. The result shows the plots for supply voltage (V_s), load current (I_l), supply current (I_s), compensator current (I_c), and DC-link voltage (V_{dc}). From $t = 0.2$ to $t = 0.4$ s, the non-linear load on the system is decreased, which leads to a corresponding increase in the DC-link voltage V_{dc}. However, the PI controller on the DC link regulates the voltage to the set reference value of 200 V. The supply voltage and supply current are in-phase relationships, indicating upf operation. This fact is also evident in Figure 4.15, which shows that the grid supplies no reactive power (blue color plot). The total three-phase load power is 4500 W and 1500 vars. Load is reduced at $t = 0.2$ s to 3000 W, 1000 vars till $t = 0.4$ s, and after that, it is regulated to the same value. The entire reactive power demand of the load is met fully by the compensator while the grid is relieved from this task. The grid meets the active power demand of the load, as shown in Figure 4.15.

Figures 4.16–4.18 show the results of an RNN controller for a three-phase system integrated into the PV source. The capacity of PV is 1 kW, and it is integrated at the DC terminals of the inverter without a boost converter stage. The inverter is designed to provide an additional active power injection capability of 1 kW along with reactive power injection. Since the three-phase connected load is higher, and the PV can only meet the load requirement partially, the rest of the power demand has to be met from the grid. At $t = 0.2$ s, two perturbations are introduced, viz. load change and PV is switched on. Both the changes are simultaneously taking place.

The load change from 4000 to 3000 W occurs at $t = 0.2$ s due to load decrease, and 1 kW PV is switched on. The effect of these variations is that grid supplies 2000 W of power and the rest is met from the PV. Load is changed at $t = 0.4$ s, and the active power demand of the load is 4500 W. It is observed that the compensator integrated to the PV source can supply only 1 kW, and the rest of 3500 W is met from the grid. Finally, no reactive power demand is met from the grid as per the designed control.

The intermediate results of the RNN algorithm are shown in Figure 4.18, which shows the load current, the estimated average value of the weight, sine template, weight of phase "a," and the error between the load current and estimated current. Since the designed controller is applied for a three-phase system, the average of all the three phase weights must be computed and filtered. The RNN control algorithm can track the fundamental component of load current suitably and performs well even under sudden load variations. The objectives of the designed controller are thus fulfilled.

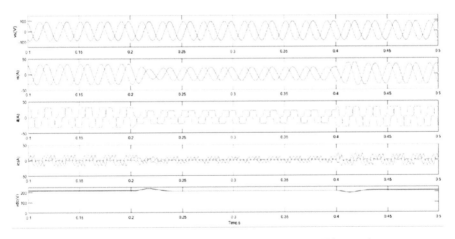

Figure 4.14   Results with RNN algorithm without PV integration.

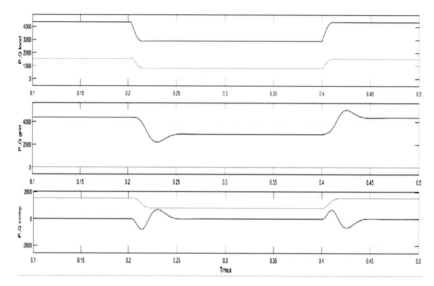

Figure 4.15   Power plots with RNN algorithm without PV integration.

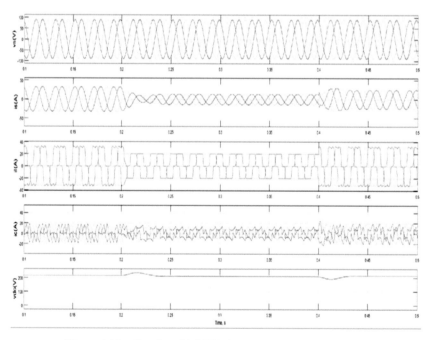

Figure 4.16   Results with RNN algorithm with PV integration.

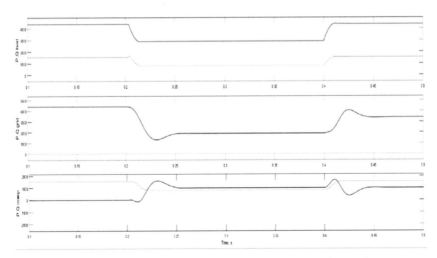

Figure 4.17   Power plots with RNN algorithm with PV integration.

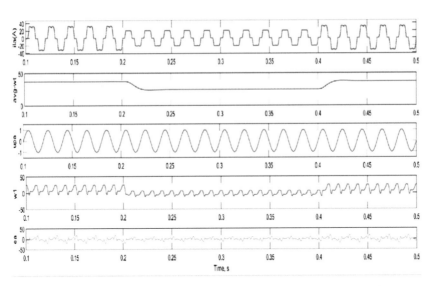

Figure 4.18   Intermediate results with RNN algorithm under load variation.

4.4 Future Scope

This chapter discusses the design of functional neural networks for the mitigation of PQ problems. Apart from T-FLANN, L-FLANN, and RNN networks, the use of ANN techniques has evolved quite a lot, especially in the last decade. This includes some newly developed NN techniques such as radial basis NN, Hopfield network, Boltzmann machine, Kohonen network, LSTM NN, and others. The NN designed and based on single/multi-layer perceptron finds immense applications in varied fields. However, all these NNs have not been effectively used for the prediction, computation, and mitigation of PQ problems in distribution systems. The use of the LSTM technique is new and encouraging, and it is based on several gates such as input/output gate, forget gate, etc. More NN techniques with the self-adaptive feature can also be designed and developed to work for achieving intelligent power systems soon. These techniques can also be designed to be more effective with lower training needs and faster convergence.

4.5 Conclusion

This chapter discusses a new class of functional expansion-based artificial neural networks, viz. T-FLANN and L-FLANN. These FLANN networks are based on expansion using trigonometric terms and Legendre functions, respectively. Further, a time-delayed recurrent neural network is also presented. All the designed controllers are made self-adaptive, and the updation of weights is achieved online via the LMS technique. The controllers can extract the fundamental component of current from the non-linear load for which these are designed. This 50-Hz current component is utilized to obtain the sinusoidal reference grid current. Both single-phase and three-phase compensators have been realized.

Moreover, the results for PV arrays integrated into these systems have also been presented. The results highlight the capability of the developed FLANN techniques in mitigating PQ problems such as harmonics, reactive power burden, load balancing in three-phase systems, and pf correction. Further, PV integrated systems have enhanced capability to provide active power support based on the installed PV source. Relevant simulation results under load and irradiance changes have been presented without and with PV integration. The results highlight the importance of designing NN-based solutions for PQ improvement.

References

[1] B. Singh, A. Chandra, K. Al-Haddad, "Power Quality: Problems and Mitigation Techniques", Wiley, 2015.

[2] B. Widrow and M. A. Lehr, "30 years of adaptive neural networks: perceptron, Madaline, and backpropagation," in Proceedings of the IEEE, vol. 78, no. 9, pp. 1415-1442, Sept. 1990, DOI: 10.1109/5.58323.

[3] P. K. Dash, H. P. Satpathy, A. C. Liew and S. Rahman, "A real-time short-term load forecasting system using functional link network," in IEEE Transactions on Power Systems, vol. 12, no. 2, pp. 675-680, May 1997, DOI: 10.1109/59.589648.

[4] D. K. Bebarta, A. K. Rout, B. Biswal and P. K. Dash, "Forecasting and classification of Indian stocks using different polynomial functional link artificial neural networks," 2012 Annual IEEE India Conference (INDICON), 2012, pp. 178-182, DOI: 10.1109/INDCON.2012.6420611.

[5] B. Singh and S. R. Arya, "Back-Propagation Control Algorithm for Power Quality Improvement Using DSTATCOM," in IEEE Transactions on Industrial Electronics, vol. 61, no. 3, pp. 1204-1212, March 2014, DOI: 10.1109/TIE.2013.2258303.

[6] A. Arora and A. Singh, "Design and analysis of functional link artificial neural network controller for shunt compensation," in IET Generation, Transmission and Distribution, no. 13, Issue 11, pp. 2280-2289, June 2019, doi:10.1049/iet-gtd.2018.6070.

[7] P. Chittora, A. Singh and M. Singh, "Chebyshev Functional Expansion Based Artificial Neural Network Controller for Shunt Compensation," in IEEE Transactions on Industrial Informatics, vol. 14, no. 9, pp. 3792-3800, Sept. 2018, DOI: 10.1109/TII.2018.2793347.

[8] A. Arora and A. Singh, "Design and Implementation of Legendre-based Neural Network in grid-connected PV systems," in IET Renewable Power Generation, no. 13, pp. 2783-2792, Nov. 2019, DOI: https://doi.org/10.1049/iet-rpg.2019.0269.

[9] H. Saxena, A. Singh, and J. N. Rai, "Application of Time Delay Recurrent Neural Network for Shunt Active Power Filter in 3-Phase Grid-tied PV System," 2019 National Power Electronics Conference (NPEC), 2019, pp. 1-6, DOI: 10.1109/NPEC47332.2019.9034769.

[10] A. Singh, M. Badoni and B. Singh, "Application of least means square algorithm to shunt compensator: An experimental investigation," 2014 IEEE International Conference on Power Electronics, Drives and Energy Systems (PEDES), 2014, pp. 1-6, doi: 10.1109/PEDES.2014.7042044.

[11] N. Roy et, "Load Forecast using ANN & VAR techniques for North Eastern Regional (NER) Grid of India," IEEE

International Conference on Power Systems ICPS, 2021, DOI: 10.1109/ICPS52420.2021.9670298.

[12] D. Wang et, "Model Predictive Control Using Artificial Neural Network for Power Converters," in IEEE Transactions on Industrial Electronics, vol. 69, no. 4, pp. 3689-3699, 2022, DOI: 10.1109/TIE.2021.3076721.

5

Application of LMS Algorithm for Mitigation of Voltage Sag as Power Quality Problem

Neelam Kassarwani

Department of Electrical Engineering, MAIT, India
E-mail: neelam.kassarwani@gmail.com

Abstract

The increased energy consumption worldwide has led to the integration of renewable power into the conventional power grid forming microgrids and smart grids as a growing necessity of the power sector forming distributed generation (DG). This trend has boosted the application of power electronic technology (PET). The augmented use of PET has revolutionized advancement in the power sector with the integration of renewable energy resources, accelerated growth of microgrids, and smart grid technologies. This has caused emergence of several power quality (PQ) issues in the electrical power systems affecting the consumers and utility as well. Mitigation of PQ issues is imperative for an uninterrupted and smooth power supply. The existing techniques using FACTS devices and custom power devices (CPDs) have employed conventional methods of their control in mitigating PQ issues of microgrids. These techniques have their limitations and give only satisfactory performance. Artificial intelligence (AI) has surfaced as a state-of-the-art trend in solving PQ issues. Numerous trends like static custom power devices in low-voltage systems and FACTS devices in flexible AC transmission systems in combination with software have stepped forward and proved their suitability in enhancing the power quality and performance of the system. In this chapter, dynamic voltage restorer (DVR) has been considered using a better technique of its control in mitigating the voltage sag problem and results have been validated using MATLAB software.

Keywords: ADAPTIVE filter, dynamic voltage restorer, power quality issues, Renewable power, voltage sag.

5.1 Introduction

Over the past few decades, industrialization of the world has encountered extensive use of electronic equipment, solid-state switching devices, nonlinear load, power electronic converters, and relaying and protective devices, which are the major sources of power quality issues. Apart from this, the increased integration of non-conventional energy sources (NCES) into the conventional power grid has reshaped the power system, resulting in a gateway to various power quality issues. The popularity of "power quality" as an incessant issue has gained substantial awareness among the utility, the consumers, the academicians, and the industries. The necessity of integrating the NCES and distributed generation (DG) systems into the main grid encourages and enhances the exploitation of solid-state technology (SST). Unfortunately, SST is the major cause of PQ issues in EPS, putting alarming financial burdens to end users. Therefore, advanced research development in the PQ analysis realm will go manifold in the years ahead with the further application of SST used in DG and NCES.

Nowadays, countries are encouraging RES-based distributed energy sources (DESs) in microgrids (MGs) with interconnected loads [1]. The power quality (PQ) issues are the significant technical challenges with the control and operation of either standalone or grid-connected MG systems due to the structure, operating mode (standalone or grid-connected), and performances of DESs in MG [2]. The increased penetration of DGs introduces numerous PQ problems such as current harmonics, voltage harmonics, voltage sag or swell, fluctuation, unbalance, malfunction of protective devices, overloading, failure of electrical equipment, etc. [3]. There is an utmost requirement for a solution to the increasing PQ problems. It is not viable to achieve a solo solution to all the existing problems.

The PQ challenges to MG, as a low-voltage network, are different from that of conventional power systems because of its unique structure, operating mode such as connected grid or standalone type, and configuration of MG distribution resources [4] such as the following:

- operating conditions of the MG DESs such as fluctuation of output power generated from RESs like PV and wind turbine (WT);

- current and voltage harmonics caused by static devices of the DESs;
- voltage and current harmonics generated by nonlinear loads in the MG system;
- unbalance in MG voltage generated by (i) unbalanced three-phase loads, and (ii) the presence of single-phase loads in the MG.

Among various PQ issues, unbalanced utility voltages, voltage sag, and voltage swell, due to a drastic increase in the integration of MGs into the main power grid, are more persistent and of great concern [5]. The voltage dip, i.e., sag, is caused by faults leading to disruption in the operation of sensitive electronic devices in DES of MG [6]. Voltage swell is another PQ issue, but its occurrence is the least.

Based on the recent PQ standards, manifold solutions have turned up in various configurations of external custom power devices (CPDs) such as distribution static compensator (DSTATCOM) [7], dynamic voltage restorers (DVR) [8], and unified PQ conditioner (UPQC) [9] to mitigate the PQ issues in the MG systems. The sagacious selection of solutions is based on the type of PQ problem that the system comes across. The DVR is substantially used as a solution to almost all types of voltage-based PQ issues such as sag/swell in MG containing PV and batteries (with excellent performance) [10], voltage fluctuations and disturbances [11], and harmonics produced by converters of MG units [12]. The shunt-connected compensating device, DSTATCOM, is used to overcome the load compensation problem and thus enhance voltage stability [13], reduce harmonics [14], mitigate unbalance [15], reduce power fluctuation [16] in MG, and increase the voltage regulation and system power factor. The hybrid compensator, UPQC, finds its use to mitigate almost all the PQ issues in the MG system through the injection or absorption of the reactive current and voltage [17]. These variants of CPD are capable of mitigating the PQ problems in the power system.

In this chapter, a series-connected static custom power device, dynamic voltage restorer (DVR), has been discussed. It is a voltage-controlled inverter that is capable of handling and mitigating almost all voltage-based utility grid PQ problems for regulating the load terminal voltage at its rated value. This device can be found in various configurations using insulated gate bipolar transistors (IGBTs), thyristors, etc., as semiconductor switches. H-bridge and three-leg are popular configurations for the practical realization of respective single-phase and three-phase dynamic voltage restorers. The series-connected CPD, dynamic voltage restorer, is the appropriate solution at the load end when efficiently controlled. These active devices, under efficient and effective control, provide reactive power compensation for the mitigation of the existing PQ issues such as voltage sag. This compensation is realized by injecting the desired voltage into the line side to regulate the load voltage to its rated value. Following are some of the PQ issues:

- voltage sag/swell in the supply line;
- unbalance in the load in a three-phase system;
- harmonics in the grid.

The performance of the solution is based on the efficiency of the control schemes in solving the PQ problems. Due to the complexity of the MG system, the PQ problems get enhanced, and the performance of the DVR is influenced by its mitigation of PQ problems. Hence, in the MG system, the mitigation of the PQ issues is to some extent only. Improved control strategies are adopted for the better performance of the devices. Numerous conventional control strategies have been reported for handling PQ issues in the MG system. These strategies have their limitations. New control strategies are developed and reported in various literature works to overcome the limitations of the control and performance of CPDs.

Control schemes play a pivotal role in the performance of compensating devices for the mitigation of PQ problems in the system. Substantial research work has been carried out in the realm of control schemes and authors have suggested and implemented several potent control techniques and algorithms to improve the performance of the CPDs. The conventional control schemes such as UT, SRF, etc., have given satisfactory performance. Therefore, these controllers need to be substituted with appropriate and improved ones [18, 19]. These new controllers are supported by artificial intelligence to make them more effective, intelligent, and adaptive. Researchers have reported several artificial-intelligence-based algorithms such as fuzzy neural networks, etc. This chapter aims at the least mean square (LMS) filtering based algorithm providing an efficient method for solving PQ issues existing in the electrical power system/MG. This scheme is adaptive in which the weights, as parameters of the adaptive filter, undergo updation process. The structure of the filter affects the performance of the CPD, computational complexity, and convergence of the algorithm. The structures like finite impulse response (FIR) filter, infinite-duration impulse response (IIR), and adaptive linear neuron or adaptive linear element (ADALINE) filter are common. In this chapter, ADALINE filter has been selected as adaptive filter because of its simple structure.

ADALINE filter is physically realized by a single-layer artificial neural network, which makes the filter adaptive. Least mean-squared (LMS) algorithm is implemented for the operation of ADALINE filter. This algorithm is a vital member of the group of stochastic gradient algorithms. As compared to other adaptive filtering algorithms, it is simple in computation and therefore widely used.

In this chapter, LMS algorithm has been selected and developed for the operation of ADALINE filter. Variants of LMS algorithms such as the basic LMS, the sign-error LMS, the NLMS sign-error, the sign-regressor LMS, the sign−sign LMS, the leaky LMS, the transform domain LMS, the robust variable step-size LMS, the block LMS, the complex LMS, the affine LMS, the complex affine LMS, and many more have been reported [20]. These adaptive algorithms are hardly applied in the control of DVR. In this chapter, DVR is capacitor-supported and emphasis is on ADALINE filter based basic LMS algorithm for the switching of VSC legs of the DVR to inject the desired compensating voltage for mitigating the voltage sag in the system.

5.2 Basic LMS Filtering-based Algorithm

The theory of the LMS algorithm is a three-stage process as follows:

- Wiener−Hopf equations;
- method of steepest descent;
- the stochastic algorithm, LMS.

5.2.1 Wiener−Hopf equations

Wiener−Hopf equations are developed for linear optimum filtering of the ADALINE filter by considering $x_1, x_2, \ldots, x_p$ as a set of p individual signals (unit vectors) applied to a corresponding set of input weights $w_1, w_2, \ldots, w_p$. The corresponding weighted signals such as $w_1 x_1, w_2 x_2, \ldots, w_p x_p$ are summed to give the output signal $y \ (= w_1 x_1 + w_2 x_2 + \ldots + w_p x_p)$. The error between the filter output y and its desired value d is minimized in a mean-squared manner to achieve the optimum set of weights of the weighted signals. Solution of Wiener−Hopf equations as the optimum set of weights $w_{01}, w_{02}, \ldots, w_{0p}$ are the optimum values of the weights $w_1, w_2, \ldots, w_p$ of the ADALINE filter. The ADALINE filter with weights as a solution of the Wiener−Hopf equation is acknowledged as the Wiener filter (by Haykin in 1991 and Widrow and Stern in 1985) as shown in Figure 5.1.

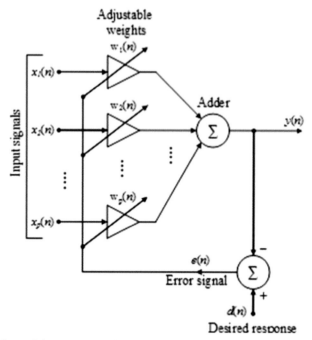

Figure 5.1 ADALINE filter to develop the Wiener–Hopf equation.

The input–output relation for this filter is given as follows:

$$y = w_1 x_1 + w_2 x_2 + \cdots + w_p x_p = \Sigma_{k=1}^{p} w_k x_k. \tag{5.1}$$

The error signal for this adaptive filter is given as follows:

$$e = d - y. \tag{5.2}$$

The error e is minimized by employing the concept of mean-squared error. The mean-squared error is the "performance measure" or "cost function" J by using the statistical expectation operator (E) as follows:

$$J = \frac{1}{2}E[e^2] = \frac{1}{2}E[(d-y)^2] = \frac{1}{2}E[(d-\Sigma_{k=1}^{p} w_k x_k)^2]. \tag{5.3}$$

Substituting eqn (5.1) and (5.2) in eqn (5.3) gives the following equation:

$$J = \frac{1}{2}E[d^2] - E[\Sigma_{k=1}^{p} w_k x_k d] + \frac{1}{2}E[\Sigma_{j=1}^{p}\Sigma_{k=1}^{p} w_j w_k x_j x_k]. \tag{5.4}$$

$$= \frac{1}{2}E[d^2] - \Sigma_{k=1}^{p} w_k E[x_k d] + \frac{1}{2}\Sigma_{j=1}^{p}\Sigma_{k=1}^{p} w_j w_k E[x_j x_k]. \tag{5.5}$$

The three expectations in eqn (5.5) are defined as follows:

- $E\left[d^2\right] = r_d$ as the mean-squared value of the desired response "d." (5.6)
- $E\left[x_k d\right] = r_{dx}$ as cross-correlation function between the desired response "d" and the set of input signals x_k where $k = 1, 2, \ldots, p$. (5.7)
- $E\left[x_j x_k\right] = r_x$ as auto-correlation function of the set of input signals where $j = 1, 2, \ldots, p$ and $k = 1, 2, \ldots, p$. (5.8)

Replacing the expectations by their respective definition representatives, eqn (5.5) gets modified as follows:

$$J = \frac{1}{2}r_d - \Sigma_{k=1}^{p} w_k r_{dx}(k) + \frac{1}{2}\Sigma_{j=1}^{p}\Sigma_{k=1}^{p} w_j w_k r_x(j,k). \quad (5.9)$$

The mean-squared error J when plotted against the weights w_1, w_2, $\ldots$, w_p takes the shape of a 3D bowl-shaped error surface as the "error-performance surface" with a global minimum point. At this point, J approaches its minimum value $J_{\min}$. The global minimum is obtained by taking the derivative of J w.r.t. w_k (the gradient $\nabla_{w_k} J$ of the error surface w.r.t. that particular weight) as follows:

$$\nabla_{w_k} J = \frac{\partial}{\partial w_k} J$$
$$= \frac{\partial}{\partial w_k}\left(\frac{1}{2}r_d - \Sigma_{k=1}^{p} w_k r_{dx}(k) + \frac{1}{2}\Sigma_{j=1}^{p}\Sigma_{k=1}^{p} w_j w_k r_x(j,k)\right).$$
$$(5.10)$$

$$= 0 - r_{dx}(k) + \sum_{j=1}^{p} w_j r_x(j,k). \quad (5.11)$$

where $k = 1, 2, \ldots, p$.

The condition for minimum error of the optimum ADALINE as an adaptive filter is

$$\nabla_{w_k} J = 0, \quad k = 1, 2, \ldots, p. \quad (5.12)$$

Taking w_{0k} as the optimum setting of weight w_k for $J_{\min}$, they are derived as the optimum weights of the filter from the set of simultaneous equations (5.10), which are the system of equations given as follows:

$$\Sigma_{j=1}^{p} w_{0j} r_x(j,k) = r_{dx}(k), \quad (5.13)$$

where $k = 1, 2, \ldots, p$.

The system of equations (5.13) is the Wiener–Hopf equations and the filter whose weights satisfy these equations is said to be the Wiener filter. The solution of these equations helps in computing the optimum weights for the ADALINE filter, which here is the Wiener filter.

5.2.2 Method of steepest descent

The solution of Wiener–Hopf equations (5.13) is the optimum weights for the ADALINE filter, which are obtained by taking the inverse of $p \times p$ matrix with different values of $r_x(j, k)$ for $j, k = 1, 2, \ldots, p$. This involves matrix inversion, which is a very complex mathematical computation. Therefore, it is suitable to replace this method with the method of steepest descent. In this method, the weights of the filter are considered as time-varying to adjust their values iteratively along the error surface and progress toward the optimum solution for the bottom or global minimum point of the error surface of the J versus w_k plot. In this hunt, the successive adjustments employed to the input weights of the filter traverse the direction of the steepest descent of the error surface which insinuates the direction opposite to that of the gradient vector. The elements of the gradient vector are given by $\nabla_{w_k} J$ for $k = 1, 2, \ldots, p$. The mean-squared error criterion for a single input weight is depicted in Figure 5.2.

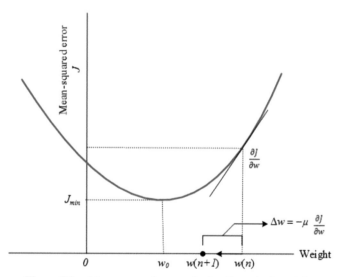

Figure 5.2   Mean-squared error criterion for a single weight.

In the method of steepest descent, the weights iteratively undergo adjustment by taking $w_k(n)$ as the value of the weight w_k of the adaptive filter computed at iteration n. The corresponding gradient of the error surface (5.11) of the filter is computed as follows:

$$\nabla_{w_k} J(n) = -r_{dx}(k) + \Sigma_{j=1}^{p} w_j(n) r_x(j,k).$$ (5.14)

The weight $w_k(n)$ at iteration n experiences an adjustment of $\Delta w_k(n)$ defined by the method of steepest descent as follows:

$$\Delta w_k(n) = -\mu \nabla_{w_k} J(n), \quad k = 1, 2, ..., p,$$ (5.15)

where μ is the learning-rate parameter and is a positive constant. Taking $w_k(n)$ as the value of the kth weight at iteration n, its updated value at iteration $(n+1)$ is computed as follows:

$$\begin{aligned} w_k(n+1) &= w_k(n) + \Delta w_k(n) \\ &= w_k(n) - \mu \nabla_{w_k} J(n), \quad k = 1, 2, ..., p. \end{aligned}$$ (5.16)

The expression (5.16) is the method of steepest descent in terms of the gradient of the error surface $\nabla_{w_k} J(n)$ for weight updation and reaching toward the solution to Wiener–Hopf equations. This method when expressed in terms of correlation functions $r_{dx}(k)$ and $r_x(j,k)$ by substituting eqn (5.14) in eqn (5.16) is

$$w_k(n+1) = w_k(n) - \mu[r_{dx}(k) - \Sigma_{j=1}^{p} w_j(n) r_x(j,k)].$$ (5.17)

The derivation of eqn (5.17) does not undergo approximations. Hence, the method of steepest descent is taken as "exact" and is based on the minimization of the mean-squared error defined in terms of iterations as follows:

$$J(n) = \frac{1}{2} E[e^2(n)].$$ (5.18)

The mean-squared error given by eqn (5.18) is the "ensemble average" computed at the nth iteration. Hence, the method of steepest descent is derived by minimizing the "sum of error squares" as eqn (5.19) by taking integration over all iterations of the algorithm for the specific realization of the adaptive filter:

$$\begin{aligned} E_{total}(n) &= \sum_{i=1}^{n} E(i) \\ &= \frac{1}{2} \sum_{i=1}^{n} e^2(i). \end{aligned}$$ (5.19)

The approach of integration in eqn (5.19) gives an identical result as in eqn (5.17), giving a new explanation of the correlation functions. This new approach of eqn (5.19) defines the auto-correlation function r_x and cross-correlation function r_{dx} as "time averages" in place of "ensemble average." The method of steepest descent functions properly when the learning-rate parameter μ is selected precisely. This method has "exact" correlation functions and suffers from the lack of knowledge of the correlation functions of the ADALINE filter when operating in an unknown environment. It is therefore suggested to replace the "exact" correlation functions, r_x and r_{dx}, by their "estimates" furnished by the least mean square (LMS) algorithm.

5.2.3 Least mean square algorithm

The LMS algorithm is established by employing instantaneous "estimates" of correlation functions $r_{dx}(k)$ and $r_x(j, k)$ in place of their corresponding "exacts" (r_{dx} and r_x) from eqn (5.7) and (5.8), respectively, which are represented by eqn (5.20) and (5.21):

$$\hat{r}_{dx}(k; n) = x_k(n)\, d(n)\,, \tag{5.20}$$

$$\hat{r}_x(j, k; n) = x_j(n)\, x_k(n). \tag{5.21}$$

Functions $r_{dx}(k)$ and $r_x(j, k)$ from eqn (5.7) and (5.8) are expressed with hats as "estimates," respectively, in eqn (5.20) and (5.21) by $\hat{r}_{dx}(k; n)$ and $\hat{r}_x(j, k; n)$ to consider weighted input signals and the desired response as a function of time. Replacing the functions $r_{dx}(k)$ and $r_x(j, k)$ by $\hat{r}_{dx}(k; n)$ and $\hat{r}_x(j, k; n)$ in eqn (5.17) modifies as follows:

$$\hat{w}_k(n+1) = \hat{w}_k(n) + \mu \left[\hat{r}_{dx}(k; n) - \sum_{j=1}^{p} \hat{w}_j(n)\, \hat{r}_x(j, k; n) \right]$$

$$= \hat{w}_k(n) + \mu \left[x_k(n)\, d(n) - \sum_{j=1}^{p} \hat{w}_j(n)\, x_j(n)\, x_k(n) \right]$$

$$= \hat{w}_k(n) + \mu \left[d(n) - \sum_{j=1}^{p} \hat{w}_j(n)\, x_j(n) \right] x_k(n)$$

$$= \hat{w}_k(n) + \mu\, [d(n) - y(n)]\, x_k(n)$$

$$= \hat{w}_k(n) + \mu e(n) x_k(n)\, k = 1, 2, ..., p\,, \tag{5.22}$$

where

$$y(n) = \Sigma_{j=1}^{p} \hat{w}_j(n) x_j(n).$$ (5.23)

Here, $y(n)$ is the output of the adaptive filter computed at iteration n in the LMS algorithm.

In the method of steepest descent, when the LMS algorithm is applied to an "unknown" environment, the weight vector $\hat{w}(n)$ follows a random trajectory rather than a precisely defined trajectory along the error surface. Due to the random trajectory of $\hat{w}(n)$ along the error surface, the LMS algorithm belongs to the "stochastic gradient algorithm" approach. As $n \to \infty$, $\hat{w}(n)$ performs Brownian motion about the optimum solution w_0 and the LMS algorithm minimizes the instantaneous estimate of the cost function $J(n)$. This makes the gradient vector "random" and its pointing accuracy "on the average" improves with an increase in the number of iterations n. The LMS algorithm minimizes the instantaneous estimate of error squared ($\mathcal{E}(n) = (1/2)e^2(n)$), which is the instantaneous estimate of the cost function $J(n)$. This reduces the storage requirement to the least with only the information of the present weights of the filter required to be stored.

The LMS algorithm implemented in adaptive filter can now operate both in the stationary and non-stationary environments of input signals (unit vectors), desired response (regulated load voltages), and the optimum solution. In the control of DVR, the LMS algorithm operates in a non-stationary environment in which input signals and desired response along with the optimum solution are time-varying. Since the task of the LMS algorithm is to keep on "seeking" and "tracking" the minimum point of the error surface, the performance of its tracking behavior can be improved by keeping the learning-rate parameter μ substantially small, but the rate of adaptation gets reduced.

5.2.4 Signal-flow graph representation of the LMS algorithm

The matrix representation of weight updation in the LMS algorithm in eqn (5.22) is given as follows:

$$\hat{w}(n+1) = \hat{w}(n) + \mu \left[d(n) - x^T(n)\hat{w}(n) \right] x(n),$$ (5.24)

where

$$\hat{w}(n) = [\hat{w}_1(n), \hat{w}_2(n), \ldots, \hat{w}_p(n)]^T,$$ (5.25)

and

$$x(n) = [x_1(n), x_2(n), \ldots, x_p(n)]^T,$$ (5.26)

Rearranging the terms in eqn (5.24) gives

$$\widehat{\boldsymbol{w}}\,(n+1) = [\boldsymbol{I} - \mu\boldsymbol{x}(n)\boldsymbol{x}^T(n)]\widehat{\boldsymbol{w}}\,(n) + \mu\boldsymbol{x}(n)d(n)\,, \qquad (5.27)$$

where $\boldsymbol{I}$ is an identity matrix of order $p \times p$. Operator z^{-1} is the unit-delay operator to imply storage and is represented as follows:

$$\widehat{w}\,(n) = z^{-1}[\widehat{\boldsymbol{w}}\,(n+1)]. \qquad (5.28)$$

Figure 5.3 is the signal-flow graph representation of the LMS algorithm in which the storage quality of eqn (5.28) is utilized and reveals its stochastic feedback system with a significant impact on its convergence behavior. Parameters constituting the feedback loop contribute to the convergence behavior representing the stability of the system. The lower feedback loop adds changeability to the behavior of the feedback loop. The statistical characteristics of the input vector $x(n)$ and the value assigned to the learning-rate parameter μ influence the convergence behavior (stability) of the LMS algorithm. Here, $x(n)$ are unit vectors (in-phase and quadrature with the source current) for generating the reference signal. Thus, in a specified environment supplying the input vector $\boldsymbol{x}(n)$, a precise selection of the learning-rate parameter μ makes the LMS algorithm convergent in the mean square if the mean-squared value of error signal $e(n)$ approaches a constant

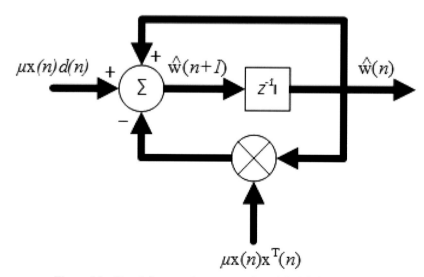

Figure 5.3   Signal-flow graph representation of the LMS algorithm.

value as the number of iterations n approaches infinity:

$$E\left[e^2(n)\right] \longrightarrow \text{constant as } n \longrightarrow \infty. \tag{5.29}$$

This method is adaptive and capable to adapt the time-varying changes in load voltages and estimate the load voltages to their reference values as the desired output. In this process, adaptive filters keep track of the dynamic nature of the reference load voltage and act as controllers to minimize the difference between the reference load voltage and the sensed load voltage by updating the weights as coefficients of the filter. This forms an adaptive control system for DVR, which repeatedly updates the parameters of the controller to reduce the error between the sensed load voltage and the reference load voltage of the EPS. Thereby, the load terminal voltage experiencing a reduction in voltage during sag dynamics can be restored and regulated.

5.3 The Adaptive LMS Filtering-based Control of DVR

The control of DVR composes of adaptive LMS filtering-based algorithm, generation of unit vectors (as input vectors), and reference voltages (as desired reference output signal) for performance study of DVR for mitigation of voltage sags in the source voltages of distribution. In this chapter, an adaptive LMS filtering-based algorithm is developed and employed in the control of capacitor-supported DVR in the system whose schematic diagram is illustrated in Figure 5.5. The MATLAB-based Simulink model of the schematic diagram presented in Figure 5.5 is shown in Figure 5.6 and the weight updation process in the basic LMS algorithm in Figure 5.7.

5.3.1 The adaptive LMS filtering-based control algorithm

LMS is an adaptive and stochastic gradient algorithm applied to find the optimal weight vector (weights) to give the least error. It is used to estimate the three-phase load voltages so that the corresponding sensed load voltages are very close to their reference values. The representation of the weight vector $\widehat{w}(n)$ and input vector $u(n)$ is shown in Figure 5.4. The weight vector is updated iteratively by implementing the updating rule given by eqn (5.22), which is represented as follows:

$$\widehat{w}(n+1) = \widehat{w}(n) + \mu e(n) u(n), \tag{5.30}$$

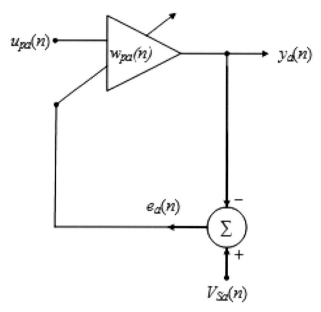

Figure 5.4   Representation of weight vector and input vector for phase "a" in-phase with the supply current i_{Sa} in DVR control.

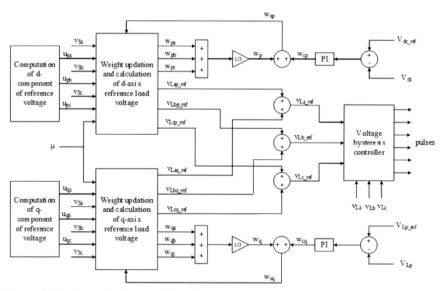

Figure 5.5   Schematic diagram of the adaptive control algorithm for the CSDVR employing LMS.

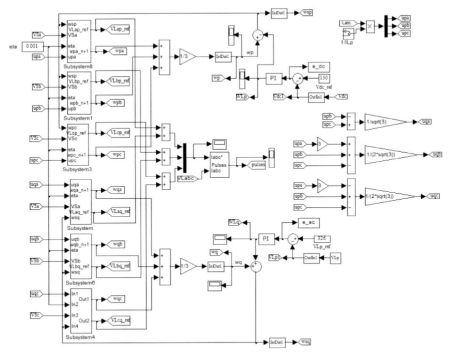

Figure 5.6 MATLAB model of DVR control using adaptive LMS filtering-based algorithm.

where, $\boldsymbol{u}(n)$ = the input vector = $\begin{bmatrix} u_a(n) & u_b(n) & u_c(n) \end{bmatrix}^T$ for the three phases, $\hat{\boldsymbol{w}}(n)$ = the present weight vector = $\begin{bmatrix} \hat{w}_a(n) & \hat{w}_b(n) & \hat{w}_c(n) \end{bmatrix}^T$ of the sensed voltage, $\hat{\boldsymbol{w}}(n+1)$ = the updated weight vector = $\begin{bmatrix} \hat{w}_a(n+1) & \hat{w}_b(n+1) & \hat{w}_c(n+1) \end{bmatrix}$, $e(n)$ = the error in the estimation = $\begin{bmatrix} e_a(n) & e_b(n) & e_c(n) \end{bmatrix}$, and μ = constant learning rate.

The learning-rate μ represents the speed with which the algorithm converges. A high value of μ worsens the stability of the algorithm, whereas a small value slows down the convergence of the algorithm. The representation of the weight vector and input vector for phase "a" is shown in Figure 5.4.

5.3.2 Generation of unit vectors

The unit vectors (u_{pa}, u_{pb}, and u_{pc}) and (u_{qa}, u_{qb}, and u_{qc}) are the input vectors respectively in-phase and quadrature with the source currents (i_{Sa}, i_{Sb}, and i_{Sc}) and utilized for the computation of corresponding fundamental active and reactive power components (w_{pa}, w_{pb}, and w_{pc}) and

(w_{qa}, w_{qb}, and w_{qc}) of the source voltages. The amplitude of the source current at PCC to compute the unit vectors in-phase and quadrature with source currents are computed as follows:

$$i_{Sp} = \left\{ \frac{2}{3} (i_{Sa}^2 + i_{Sb}^2 + i_{Sc}^2) \right\}^{1/2} \tag{5.31}$$

The unit vectors, in-phase with the source currents (i_{Sa}, i_{Sb}, and i_{Sc}), are computed as given by

$$\begin{aligned}
u_{pa} &= i_{Sa}/i_{Sp}, \\
u_{pb} &= i_{Sb}/i_{Sp}, \text{ and} \\
u_{pc} &= i_{Sc}/i_{Sp}
\end{aligned} \tag{5.32}$$

The unit vectors in quadrature with three-phase PCC load currents (i_{Sa}, i_{Sb}, and i_{Sc}) are computed utilizing the in-phase unit vectors in eqn (5.32), which are given as follows:

$$\begin{aligned}
u_{qa} &= -u_{pb}/\sqrt{3} + u_{pc}/\sqrt{3} \\
u_{qb} &= \sqrt{3}u_{pa}/2 + (u_{pb} - u_{pc})/2\sqrt{3} \\
u_{qc} &= -\sqrt{3}u_{pa}/2 + (u_{pb} - u_{pc})/2\sqrt{3}
\end{aligned} \tag{5.33}$$

5.3.3 Estimation of reference load voltages

The estimation process of fundamental active and reactive power components of the reference load voltages utilizing eqn (5.30) respectively in-phase and quadrature components of unit vectors of the source currents in eqn (5.32) and (5.33) is illustrated in Figure 5.6. The sensed DC-link voltage of VSC of DVR is fed to a low-pass filter (LPF) to remove the ripple components. The reference DC-link voltage v_{DC_ref} and filter-sensed DC-link voltage v_{DC} of the VSC of DVR are compared in the summer block and error e_{DC} is realized over the PI controller, forming a DC-link control loop. The output of this PI controller is the loss component w_{Lp} (= v_{loss}) due to the switching of the three-leg VSC of the DVR. The loss component w_{Lp} computed at $(n+1)$th sampling instant is given by

$$\begin{aligned}
w_{Lp}(n+1) = w_{Lp}(n) + &K_{pd} \{e_{DC}(n+1) - e_{DC}(n)\} \\
&+ K_{id}(e_{DC}(n+1))
\end{aligned} \tag{5.34}$$

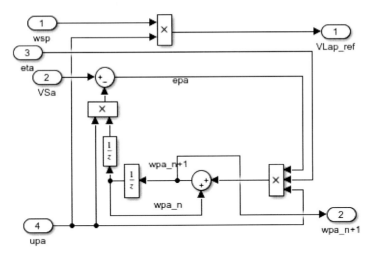

Figure 5.7 Basic LMS updation process.

where K_{pd} and K_{id} are gain parameters of PI controller realized over filtered DC-link voltage; $e_{DC}(n)$ and $e_{DC}(n + 1)$ are, respectively, errors at nth and $(n + 1)$th instants; and $w_{Lp}(n)$ and $w_{Lp}(n + 1)$ are corresponding amplitudes of loss component injected by DVR in-phase with the source current at nth and $(n + 1)$th instants.

In this work, a filtering-based LMS algorithm is developed to estimate and extract the updated weights of the fundamental active power component of load voltage [21]. Figure 5.7 shows the estimation of the active power component of phase "a" of the source voltage, which is presented below showing the updation process in the ADALINE filter:

$$w_{pa}(n + 1) = w_{pa}(n) + \mu \times e_{pa}(n) \times u_{pa}(n), \qquad (5.35)$$

where

$$e_{pa}(n) = v_{Sa}(n) - w_{pa}(n - 1) \times u_{pa}(n) \qquad (5.36)$$

is the error component of the proposed control algorithm extracted for phase "a" and $v_{Sa}(n)$ is the PCC voltage of phase "a" at the nth sampling instant; $w_{pa}(n - 1)$ is the amplitude of the active power component of source voltage (v_{Sa}) at the $(n - 1)$th sampling instant.

The updation processes in phases "b" and "c" of source voltages are computed by following the same pattern for the corresponding proposed LMS algorithms. The rest of the estimation and computation is done by considering the basic LMS algorithm.

The learning-rate μ accounts for updation of error components. In this thesis, the value of μ is selected to be 0.001and is depicted in Figure 5.6. The estimation of active power components for phases"b" and "c" of source voltages are computed as given by eqn (5.37) and (5.38), respectively:

$$w_{pb}\left(n+1\right)=w_{pb}\left(n\right)+\mu\times u_{pb}(n)\times e_{pb}(n) \tag{5.37}$$

$$w_{pc}\left(n+1\right)=w_{pc}\left(n\right)+\mu\times u_{pc}(n)\times e_{pc}(n) \tag{5.38}$$

The average magnitude of w_{pa}, w_{pb}, and w_{pc} is computed to give the weighted value of the fundamental active power components of reference load voltage as given by eqn (5.39):

$$w_{p}=\left(w_{pa}+w_{pb}+w_{pc}\right)/3 \tag{5.39}$$

The loss component of DVR (w_{cp}) with fundamental active power components of reference load voltage (w_{p}) gives the total weighted value of the fundamental active power component of reference load voltage as in the following equation:

$$w_{sp}=w_{p}-w_{Lp} \tag{5.40}$$

The weighted value of the fundamental active power component of the reference load voltage, in phase with unit vectors of the source current eqn (5.38) is the estimated in-phase component of reference load voltage as follows:

$$v_{\text{Lap\_ref}}=w_{sp}\times u_{pa};\ v_{\text{Lbp\_ref}}=w_{sp}\times u_{pb};\ v_{\text{Lcp\_ref}}=w_{sp}\times u_{pc} \tag{5.41}$$

Similarly, weights of reactive power component of three-phase load voltages "a," "b," and "c" are computed in the following equation:

$$w_{qa}\left(n+1\right)=w_{qa}\left(n\right)+\mu\times u_{qa}(n)\times e_{qa}(n)$$
$$w_{qb}\left(n+1\right)=w_{qb}\left(n\right)+\mu\times u_{qb}(n)\times e_{qb}(n)$$
$$w_{qc}\left(n+1\right)=w_{qc}\left(n\right)+\mu\times u_{qc}\left(n\right)\times e_{qc}\left(n\right) \tag{5.42}$$

where μ is the learning rate of the proposed control algorithm for reactive power component estimation and has the same value as that for active power component estimation, which is 0.001. The average of w_{qa}, w_{qb}, and w_{qc} computes the average magnitude of weighted value of the fundamental reactive power components of reference load voltage as follows:

$$w_{q}=\left(w_{qa}+w_{qb}+w_{qc}\right)/3 \tag{5.43}$$

The voltage error between the amplitude of the reference value ($v_{\text{Lp\_ref}}$) of load voltage and amplitude of the sensed load voltage (v_{Lp}) is realized over PI controller forming an AC-link control loop. The AC-link PI controller output (w_{Lq}) is the reactive power component injected by DVR to regulate load terminal voltages to the desired value to mitigate sag in the source voltages. The updated output of the AC-link PI controller, represented by $w_{Lq}(n+1)$ at the $(n+1)$th sampling instant, is computed as given in the following equation:

$$w_{Lq}(n+1) = w_{Lq}(n) + K_{pq} \times \{e_{AC}(n+1) - e_{AC}(n)\}$$
$$+ K_{iq} \times e_{AC}(n+1) \tag{5.44}$$

where K_{pq} and K_{iq} are gain parameter PI controller; $e_{AC}(n)$ and $e_{AC}(n+1)$, respectively, are the errors between reference load voltage and sensed load voltage at nth and $(n+1)$th instants; and

$w_{Lq}(n)$ and $w_{Lq}(n+1)$, respectively, are the corresponding amplitudes of reactive power components injected by DVR at nth and $(n+1)$th instants.

The reactive power component w_q with w_{Lq} gives the total weighted value of the fundamental reactive power component of reference load voltage given as follows:

$$w_{sq} = w_{Lq} + w_q \tag{5.45}$$

The weighted value of the fundamental reactive power component of reference load voltage, in quadrature with the unit vectors of load current (5.48), is the estimated quadrature component of reference load voltages as in the following equation:

$$v_{\text{Laq\_ref}} = w_{sq} \times u_{qa}$$
$$v_{\text{Lbq\_ref}} = w_{sq} \times u_{qb}$$
$$v_{\text{Lcq\_ref}} = w_{sq} \times u_{qc} \tag{5.46}$$

The total reference load voltages ($v_{\text{La\_ref}}$, $v_{\text{Lb\_ref}}$, and $v_{\text{Lc\_ref}}$) are computed as the sum of corresponding in-phase (5.41) and quadrature (5.46) components. Errors in load voltage computed as the difference of sensed load voltages (v_{La}, v_{Lb}, and v_{Lc}) and reference load voltages ($v_{\text{La\_ref}}$, $v_{\text{Lb\_ref}}$, and $v_{\text{Lc\_ref}}$) are realized over hysteresis controller (HC) to generate relevant six switching pulses for indirect voltage control by generating DVR side compensating voltages to be injected into a three-phase distribution system. The indirect voltage control extracts sinusoidal reference load voltages. The simulation experiment has been performed and the performance of DVR is studied by employing the above proposed LMS algorithms.

Results obtained from the simulation experiments are presented to study the performance of DVR with the proposed LMS algorithms for the mitigation of sag in the source voltages of the CSDVR-connected system.

5.4 Results and Performance Study

The above adaptive control scheme is applied for the mitigation of voltage sag and performance is studied through simulation experiments under the simulation experiment environment listed in Table 5.1 for different cases of voltage sag conditions (Table 5.2) experienced in the source voltages of three-phase CSDVRCS.

Table 5.1   Simulation experiment environment for the performance study of DVR.

S. no.	Study environment	Remarks	
1.	Sag conditions	Refer Table 5.2	
2.	Time of simulation run	0–2.0 s	
3.	Duration of sag	0.5 s	
4.	Simulation time for which results are considered	0.6–2.0 s	
5.	Simulation time at which sag is initiated in the source voltage	0.9 s	
6.	Simulation time by which sag terminates	1.4 s	
7.	Performance indices considered	1.2-norm of the error	4. RMSE
		2. MSE	5. Absolute minimum error
		3. ISE	6. Absolute maximum error

Table 5.2   Cases of different voltage sag conditions.

S. no.	Case	Sag condition	Sag in phases		
			a	*b*	*c*
1.	Case I	Balanced	30%	30%	30%
2.	Case II	Unbalanced	30%	20%	10%
3.	Case III	Two-phase	30%	30%	0%
4.	Case IV	One-phase	30%	0%	0%

The simulation results are presented as follows.

5.4.1 Waveform plots for the study of the dynamic performance of DVR

The waveforms of three-phase PCC voltage (v_{Sabc}), load voltage (v_{Labc}), load current (i_{Labc}), the amplitude of PCC voltage (V_{Sp}), the amplitude of load voltage (V_{Lp}), VSC-side DVR voltage (V_{VSC}), line-side voltage injected by DVR (v_{DVR}), and DC-link voltage (V_{dc}) are shown in Figure 5.8(a) to analyze its dynamic performance:

Case I: Under balanced voltage sag condition
Case II: Under unbalanced voltage sag condition
Case III: Under two-phase voltage sag condition
Case IV: Under one-phase voltage sag condition

It has been observed that under all the voltage sag conditions, DVR has injected the required compensating voltage in series with the feeder line, resulting in regulation of the DC-link and AC-link (load) voltages to their

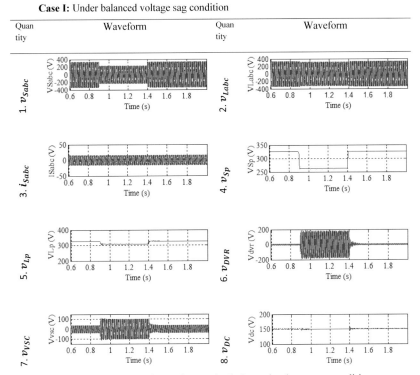

Figure 5.8(a)   Waveform plots under balanced voltage sag condition.

Case II: Under unbalanced voltage sag condition

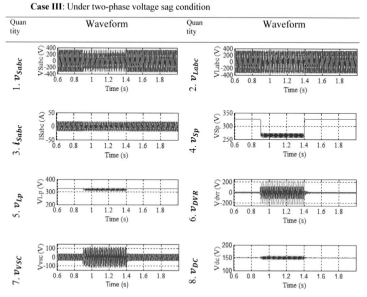

Figure 5.8(b)   Waveform plots under unbalanced voltage sag condition.

Case III: Under two-phase voltage sag condition

Figure 5.8(c)   Waveform plots under two-phase sag condition.

Case IV: Under one-phase voltage sag condition

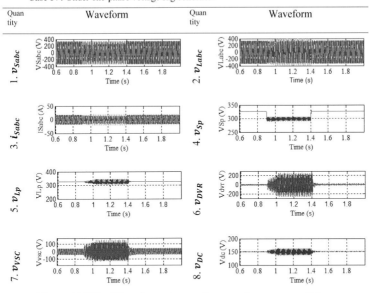

Figure 5.8(d) Waveform plots under one-phase sag condition.

rated voltages without experiencing undershoot or overshoot during sag dynamics.

5.4.2 Error plots for the study of the dynamic performance of DVR

Error plots of e_{DC} and e_{AC} are presented in Figure 5.9.

From the above figure, it is observed that errors in DC-link and AC-link feedback loops are quite reduced.

5.4.3 Behavior of the pattern of fundamental active and reactive power components

The behaviors of the pattern of components w_p, w_{sp}, and w_{Lp} and w_q, w_{sq}, and w_{Lq} during the simulation run are shown in Figures 5.10(a) and 5.10(b) under the simulation experiment environment. Fundamental active power component (w_{sp}) and reactive power component (w_{sq}) are the weights utilized for the estimation of reference load voltages. Other components, w_p, and w_{Lp} as active power components and w_q, and w_{Lq} as reactive power components, estimate the corresponding weights w_{sp} and w_{sq}. It is observed that

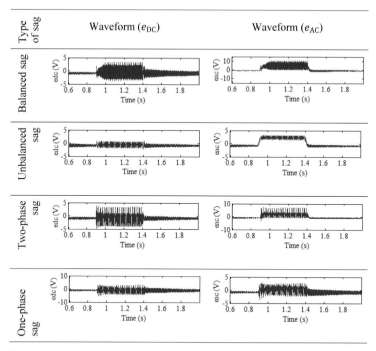

Figure 5.9   Plots of errors e_{AC} and e_{DC} using basic LMS control algorithm.

the behavior of the pattern of all active and reactive power components is as required and the proposed algorithm has estimated the reference load voltages for switching of IGBTs. This has rendered DVR to inject the required voltage to regulate DC-link and AC-link voltages. The proposed control algorithm is found to be fast and precise.

5.4.4 Computed data from error plots study of the steady-state performance of DVR

Performance indices, in terms of errors, respectively, in v_{DC} and v_{Lp}, are the computed values of 2-norm of error, MSE, ISE, RMSE, maximum of absolute error, and a minimum of absolute error from the corresponding error plots of e_{DC} and e_{AC} and are tabulated in Tables 5.3(a) and 5.3(b). These values are computed under different cases of voltage sag conditions (Table 5.3) in the three-phase source voltages of capacitor-supported DVR-connected system.

It is observed that all the computed values, as given in Tables 5.3(a) and 5.3(b) for error performance in v_{DC} and v_{Lp}, have nominal values showing

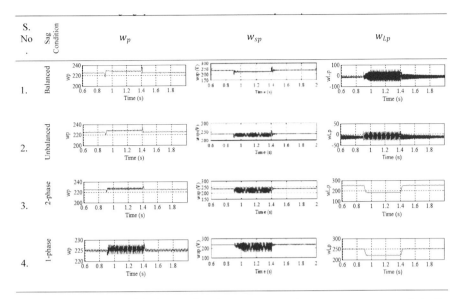

Figure 5.10(a) Plots of fundamental active w_p, w_{sp}, and w_{Lp} power components in the basic LMS algorithm.

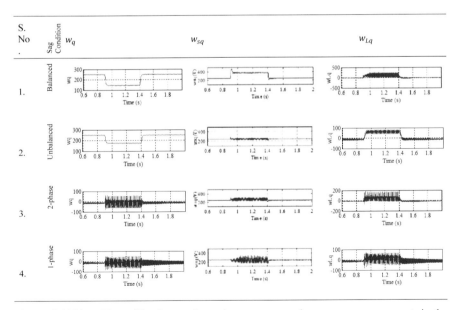

Figure 5.10(b) Plots of fundamental reactive w_q, w_{sq}, and w_{Lq} power components in the basic LMS algorithm.

Table 5.3(a) Performance indices in v_{DC} using the basic LMS algorithm.

S no.	Type of sag condition	2-Norm (×e+003)	MSE	ISE (×e+006)	RMSE	max(abs(e_{DC}))	min(abs(e_{DC})) (×e-006)
1	Balanced sag	0.55375	0.8760	0.3066	0.9360	6.6116	5.4796
2	Unbalanced sag	0.82871	1.9907	0.6868	1.4109	4.3439	85.1220
3	Two-phase sag	1.4647	6.2183	2.1453	2.4937	6.9292	169.6400
4	One-phase sag	2.5570	18.9550	6.5395	4.3537	13.0054	21.627

Table 5.3(b) Performance indices in v_{Lp} using the basic LMS algorithm.

S. no.	Type of sag condition	2-Norm (×e+003)	MSE	ISE (×e+006)	RMSE	max(abs(e_{AC}))	min(abs(e_{AC})) (×e-006)
1	Balanced sag	5.5610	89.6375	30.9250	9.4677	22.4381	177.6600
2	Unbalanced sag	2.0243	11.7076	4.0977	3.4216	10.4899	11.0610
3	Two-phase sag	3.1824	28.9359	10.1280	5.3792	16.5214	287.2300
4	One-phase sag	3.6250	37.5454	13.1410	6.1274	21.0248	292.6500

better regulation of DC-link voltage v_{DC} and AC-link voltage v_{Lp} with minimum error and rise time during sag dynamics. This ensures a better steady-state performance of DVR.

5.5 Discussion

Basic LMS algorithms with PI controllers, each in DC-link and AC-link feedback control loops, are developed and employed for adaptive LMS filtering-based control of DVR. This adaptive algorithm has successfully computed the desired reference load voltage with precise switching of VSC. This enabled DVR to inject the desired compensating voltages, in-phase and quadrature with the feeder line and mitigated the voltage sags experienced in the source voltages of the capacitor-supported DVR-connected system. This algorithm has also eliminated the drawbacks of undershoot and overshoot in the restored and regulated voltages of DC-link and AC-link of DVR, which is normally observed.

The only drawback with the proposed LMS filtered-based control algorithms, developed and employed for the control of DVR to mitigate the voltage sags in the source voltages of the capacitor-supported DVR-connected system, is that the regulated voltages experience oscillations.

5.6 Conclusion

DVR has given excellent performance in mitigating voltage sags in source voltages of capacitor-supported DVR-connected systems, thereby restoring and regulating load terminal voltages. Since undershoot and overshoot get introduced in the regulated load terminal voltages with the conventional control schemes, it had been a big challenge for the devices connected to the system. The adaptive LMS filtering-based algorithm with variant basic was implemented for the control of DVR for sag mitigation. Sag condition from the worst to the most common cases was considered in the performance analysis of DVR. The performance of DVR is excellent with no undershoot and overshoot. Oscillations are observed in the regulated load voltage in the waveform plots and its elimination is acknowledged.

References

[1] M. S. Nazar, A. R. Sadegh, and A. Heidari, "Optimal Microgrid Operational Planning Considering Distributed Energy Resources," in Microgrid Architectures, Control and Protection Methods. 3rd ed., Western Electric Co., Springer 2020, pp. 491-507.

[2] R. M. Kamel, "New inverter control for balancing standalone microgrid phase voltages: A review on MG power quality improvement," Renewable and Sustainable Energy Reviews, vol. 63, pp. 520-532, Sep. 2016.

[3] P. K. Ray, S. R. Mohanty, N. Kishor, and J. P. Catalão, "Optimal feature and decision tree-based classification of power quality disturbances in distributed generation systems," IEEE Transactions on Sustainable Energy, vol. 5, no 1, pp. 200-208, 2013.

[4] O. Palizban, K. Kauhaniemi, and J. M. Guerrero, "Microgrids in active network management–part II: System operation, power quality and protection," Renewable and Sustainable Energy Reviews, vol. 36, pp. 440-451, Aug. 2014.

[5] Á. Espín-Delgado, S. Rönnberg, T. Busatto, V. Ravindran, and M. Bollen, "Summation law for supraharmonic currents (2–150 kHz) in low-voltage installations," Electric Power Systems Research, vol. 184, pp. 106325, July 2020

[6] M. Gayatri, A. M. Parimi, and A. P. Kumar, "Utilization of Unified Power Quality Conditioner for voltage sag/swell mitigation in microgrid," presented at Biennial International Conference on Power and Energy Systems: Towards Sustainable Energy (PESTSE), 2016, pp. 1-6.

[7] L. Naik and K. Palanisamy, "Design and Performance of a PV-STATCOM for Enhancement of Power Quality in Micro Grid Applications," International Journal of Power Electronics and Drive Systems (IJPEDS), vol. 8, no 3, pp. 1408-1415, Sep. 2017.

[8] T. HS and T. R. D. Prakash, "Reduction of Power Quality Issues in Micro-Grid Using Fuzzy Logic Based DVR," International Journal of Applied Engineering Research, vol. 13, no.11, pp. 9746-9751, 2018.

[9] M. Gayatri, A. M. Parimi, and A. P. Kumar, "Utilization of Unified Power Quality Conditioner for voltage sag/swell mitigation in microgrid," presented at Biennial International Conference on Power and Energy Systems: Towards Sustainable Energy (PESTSE), 2016, pp. 1-6.

[10] X. Han, R. Cheng, P. Wang, and Y. Jia, "Advanced dynamic voltage restorer to improve power quality in microgrid," presented at IEEE Power & Energy Society General Meeting, 2013, pp. 1-5.

[11] M. A. Mansor, M. M. Othman, I. Musirin, and S. Z. M. Noor, "Dynamic voltage restorer (DVR) in a complex voltage disturbance compensation," International Journal of Power Electronics and Drive Systems, vol. 10, no.4, p. 2222, 2019.

[12] P. Kishore and D. N. Kumar, "Performance and Comparison of Harmonics Using Active Power Filters and DVR in Low-Voltage Distributed Networks," in Innovations in Electrical and Electronics Engineering, 2nd ed., Western Electric Co., Springer 2020, pp. 291-303.

[13] F. Hamoud, M. L. Doumbia, and A. Chériti, "Voltage sag and swell mitigation using D-STATCOM in renewable energy based distributed generation systems," presented at Twelfth International Conference on Ecological Vehicles And Renewable Energies (EVER), 2017, pp. 1-6.

[14] N. P. Lakshman and K. Palanisamy, "Design and Performance of a PV-STATCOM for Enhancement of Power Quality in Micro Grid Applications," International Journal of Power Electronics and Drive Systems, vol. 8, no.3, p. 1416, 2017.

[15] L. Wang, W.-S. Liu, C.-C. Yeh, C.-H. Yu, X.-Y. Lu, B.-L. Kuan, et al., "Reduction of three-phase voltage unbalance subject to special winding connections of two single-phase distribution transformers of a microgrid system using a designed D-STATCOM controller," IEEE Transactions on Industry Applications, vol. 54, pp. 2002-2011, 2017.

[16] L. E. Christian, L. M. Putranto, and S. P. Hadi, "Design of Microgrid with Distribution Static Synchronous Compensator (D-STATCOM) for Regulating the Voltage Fluctuation," presented at IEEE 7th

International Conference on Smart Energy Grid Engineering (SEGE), 2019, pp. 48-52.

[17] M. Gayatri, A. M. Parimi, and A. P. Kumar, "Utilization of Unified Power Quality Conditioner for voltage sag/swell mitigation in micro-grid," presented at Biennial International Conference on Power and Energy Systems: Towards Sustainable Energy (PESTSE), 2016, pp. 1-6.

[18] Digital Control Systems Implementation and Computational Techniques; Zhiqiang Gao, in Control and Dynamic Systems, 1996.

[19] I. D. Landau, in Encyclopedia of Physical Science and Technology (Third Edition), 2003 I Adaptive Control Systems: Basic Principles

[20] A. D. Poularikas, Adaptive filtering: Fundamental of Least Mean Squares with MATLAB, 1st ed. Boca Raton: CRC Press, Taylor & Francis Group, 2015. 162M. H. Hayes, Statistical Digital Signal Processing and Modelling, New York: John Wiley & Sons, Inc., 1996.

[21] S. Haykin and B. Widrow, Least-Mean-Square Adaptive Filters, New Jersey, USA: John Wiley & Sons, 2003.

6

Blockchain based Solution for Electricity Supply Chain in Smart Grids

Devanjali Relan[1], Kiran Khatter[1], and Neelu Nagpal[2]

[1]BML Munjal University, Gurgaon, Haryana
[2]Maharaja Agrasen Institute of Technology, Rohini, Delhi
E-mail: devanjali.relan@bmu.edu.in; kiran.khatter@bmu.edu.in;
nagpalneelu1971@ieee.org

Abstract

Blockchain technology is an immutable digital ledger of transactions that can be shared across a network of computers to make the process of recording transactions and tracking assets more efficient. This emerging technology drew considerable interest in smart grid and energy sectors. With the expansion of the industrial age, there is a great demand for electricity in the present system. The smart grid concept was offered to efficiently distribute electricity with minimal losses and a high level of supply security. This concept helped to convert energy consumers into producers (such as excess energy produced by solar panels) who can sell excess power back to the grid. This approach, however, adds to the complexity of the existing system, such as how a transaction between producers and consumers is carried out, authorised, and documented. The introduction of blockchain technology in the electricity supply chain guarantees transparent and secure systems. This chapter describes how blockchain can be integrated into smart grid architecture to supply the electricity in a transparent, authenticated, secure and reliable system. The proposed decentralised and distributed solution aims to preserve the supply chain ecosystem in the smart grid.

6.1 Introduction

For years, the generation of electricity was mainly based on fossil fuel and the resources to do that was very expensive, especially for small scale use. We are now on our way to a future where more households produce their electricity. We can also envision a future with smart grids, [1] in which power networks are more than just a static infrastructure and electricity supply. As a result, the smart grid may be conceived of as a next-generation power infrastructure that creates a distributed automated energy delivery system using two-way data and electricity flows [2]. To this end this chapter contribute towards and explains:

- the state of art method in the related field
- how blockchain may be integrated into smart grid architecture to ensure that electricity is delivered in a transparent, authenticated and secure manner.
- the proposed decentralised and distributed system that attempts to protect the smart grid's supply chain ecology.

6.1.1 Smart grid

A smart grid is a network that allows electricity and data to flow in both directions, using smart metering as a first step. It is a two-way network of information sharing between electricity grids and users that is both transparent and instantaneous [3]. In this model, consumers pay for the electricity they buy from power plants based on their consumption. With the advancement in technology, automation in electricity generation has led to mass production. This automation has revolutionised the way people create and distribute electricity, in addition to lowering the cost of power generation [4]. When a consumer generates more electricity than they consume, they sell it to others, so acting as alternative energy sources [4, 5]. By selling excess energy to neighbouring customers or the grid, these consumers who harvest renewable energy sources become producers-cum-consumers (prosumers). This approach encourages consumers to use renewable energy sources [6]. Several variables are driving the move from traditional electric grids to smart grids, such as deregulation of the energy market, evolution in metering, decentralisation (distributed energy) etc. [7].

The fundamental distinction between the smart grid and the traditional power grid is that the smart grid is decentralised, with numerous small power producers compared to the traditional power grid's centralised huge power

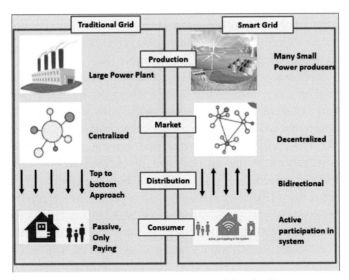

Figure 6.1   Smart grid vs. traditional grid

plants. The two-way flow of electricity and data enabled by a smart grid allows information and data to be fed to various stakeholders in the electricity market, which can then be investigated to enhance the grid, predict potential issues, respond more quickly when challenges arise, and build new functions and services as the power situation changes. The difference between smart grid and the conventional grid is shown in Fig. 6.1.

6.1.2 Challenge

The distributed energy sources, as compared to traditional power-producing methods, have unique characteristics that pose significant problems to the present distribution system [8]. The electricity grid's primary function is to transport energy in a safe and reliable manner [9]. Smart grids present a solution for integrating these distributed energy sources into the power grid's security of supply.

Smart grids are designed to make local production and consumption easier for prosumers (a new type of grid user which produces and consumes electrical energy)and consumers [10], and this decreases the transmission losses by increasing local energy production and consumption. This system will allow to trade electricity with one another in a peer-to-peer manner. To centrally handle electricity trading and transaction between Prosumers and consumers on a peer-to-peer basis would be prohibitively expensive and would necessitate a complicated communication infrastructure[11]. Thus, it is evident that a decentralised approach is preferable [12].

Smart grid solutions, while their many benefits, add complexity to the existing electrical industry, particularly at the transaction level. For example, as there are many consumers and generators in the network so it becomes a question of concern that who will validate the transaction, i.e. if the consumer has paid for the electricity or not, so that the generator will deliver the electricity [5]. Advances in the internet and blockchain technology have created a distributed accounting system that is both tamper-resistant and extremely trustworthy [13]. Blockchain-based applications could provide solutions to challenges of varying degrees of complexity. Blockchain can not only enhance the security of the grid system, but it can aid in the realisation of the efficient and reliable distributed smart grid system [14, 15].

6.1.3 Electric power supply chain management

The main aim of the smart grid is to not only distribute energy to the users but also to achieve consumption efficiency by involving the users in decision making. It integrates the actions of various stakeholders such as producers, distributors, prosumers, residents, commercial users and industrial users to efficiently distribute economic energy supplies (Figure 6.2).

In this system, there may be few consumers who consume the energy from various sources and also produce energy to give it back to the grid, which can be used during peak hours. Through the cumulative effort of various stakeholders in the supply chain, the end product reaches to consumer. Thus supply chain management (SCM) integrates the business processes that involve stakeholders such as manufacturers, suppliers and end-users to offer products and services with the aim to provide value to the consumer and achieve a competitive advantage [16]. Electric Power Supply Chain (EP-SC) Management is the integration of essential operations such as generation, transmission, distribution, and sales to maintain supply and demand balance and provide power and related services that bring value to producers, distributors, prosumers and consumers (Refer Fig. 6.3).

In EP-SCM, the smart grid supplies energy with the help of a power plant that is equipped with specialized software, hardware and automation technologies. Power plant acts as a producer which delivers further benefit to micro-grids. Micro-grids distribute energy to various consumers such as household, commercial and industrial. There are few consumers who have an

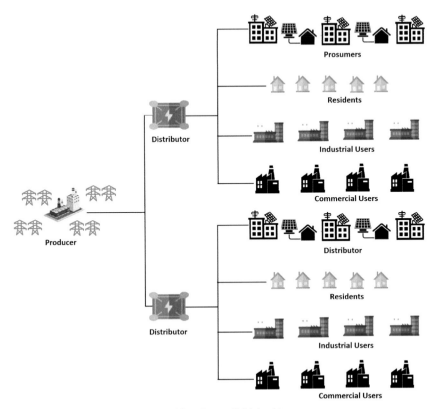

Figure 6.2 Smart Grid Architecture

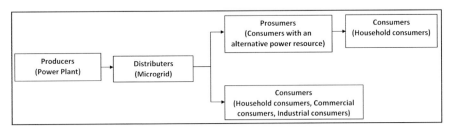

Figure 6.3 Different Stakeholders in Electrical Supply chain

alternative source of energy and act as prosumers and further distribute the energy to household consumers.

The supply chain faces certain problems, such as inconsistency in information, sudden changes in lead times and price, and shortage of material if there is a lack of visibility and transparency in the system. This inefficient supply chain can intensify a trust issue among the stakeholders, and it requires a mechanism to improve information exchange, and verifiability [17].

The main goal of this paper is to explain how electricity is delivered in a transparent, authenticated, secure, and reliable manner using blockchain in the smart grid architecture. We proposed a decentralised and distributed method to keep the supply chain ecosystem in the smart grid intact.

6.2 Related Work

The complex supply chain in the smart grid is an association of organization, human resources, and information [18]. The incompetent supply chains require better information sharing and verifiability; otherwise, it may result in a trust issue [19]. The power industry is being transformed by smart grid technology and changing every aspect of the grid supply chain [20], but at the same time, it introduces new security and intelligence challenges [21].

Data exchange in supply chain management could become safer and more transparent thanks to emerging blockchain technology [22]. The smart grid supply chain may ensure secure transaction data storage while also encouraging the use of green energy and promoting sustainable development by utilizing blockchain technology [23].

Blockchain is a decentralized database system that is open and transparent. In blockchain technology, the Consensus Algorithm (CA) is a vital and inseparable component. The availableness of a diverse set of consensus algorithms leads to the development of blockchain in various applications [24]. Sylim Partick et al. [25, 26] created a distributed application using Swarm as a Distributed File System (DIS) and smart contracts.

Huang Yan et al. proposed Drugledger, a scenario-based blockchain system for drug tracking and regulation [27]. In their service architecture, they segregated the service provider into three discrete service components to assure the authenticity and privacy of traceable data. To create a peer-to-peer network, the communication module employs gossip. The consensus module is implemented using the Algorand cryptocurrency and its algorithm. Drugledger and Algorand users are weighted based on the number of legitimate transactions they've completed in the past and based on the balance they own, respectively.

The application of blockchain technology to the creation of smart grids has piqued the interest of academics. Many academics have proposed blockchain-based solutions to overcome the security and intelligence challenges of smart grids. A sovereign blockchain was utilized by Gao et al. [28] to automatically record and control the power in a power system. Furthermore, blockchain-based smart grid solutions were proposed by Mengelkamp et al. [29], Wang et al. [30], Jamil et al. [31] and Pop et al. [32]. Integration of supply chains with blockchain technology has the potential to save time, energy, and money for all of the energy companies involved. Moreover, through traceability, transparency, and tradeability, blockchain can improve the consumer experience in EP-SCM.

6.3 Background of Blockchain

Blockchain is a decentralized and distributed ledger that helps to verify the provenance of digital assets. These digital assets are stored in the form of blocks in the blockchain using crypto-hashing, which makes them unalterable. The transactions related to each block are stored in the blockchain, and whenever a new transaction occurs, its timestamp is permanently recorded in the blockchain. The blocks stored in the chain are connected to each other in the form of a linked list. If any new block is added in the blockchain, its timestamp, nonce, Merkle root, and a hash of the previous block are stored in the block header. Since the hash of a particular is calculated by hashing the information stored in the block and block header, a change in the information of the block will cause a change in the hash of that block itself and the blocks after it. If an intruder makes changes in one block, it will affect the hash of other blocks. This would be sensed and disapproved by all nodes on a distributed network as the copy of the chain what they have is different. Thus tampering in one block requires a change in every block stored in the blockchain, which is not possible. Thus information stored in the blockchain is shared among different parties in a secure manner which brings trustworthiness. [33].

- **Distributed Ledger**
  Distributed Ledger Technology (DLT) securely stores information and allows concurrent access of information that is replicated across multiple locations of the network. Authorized users access information using keys and cryptography signatures. There is no central node needed to administer functionality and track changes in the database. In a distributed network, different nodes can communicate with each other without the presence of any centralized or intermediary node. For every change, a consensus has to be obtained from all other nodes from the

network. The information stored in the network becomes immutable and is regulated by the rules of the network. The information stored in a distributed database is consistent across multiple locations, and each update is recorded by all the nodes in the network. If there is any change in the information or any new addition is made to the database, it is copied to all the nodes of different locations of a distributed network. Thus in a distributed network, a single point of failure is removed, and transparency is increased.

- **Transaction**
Whenever there is an exchange of data in the form of digital assets or money happens among different nodes of the blockchain network, it is called a transaction.

- **Distribution**
Once a transaction is authenticated, a block for that transaction is created and sent across the chain. Further transactions are validated by miners by using proof of work. In order to do this, miners solve a mathematical puzzle to obtain the nonce value which is required to generate a hash for a particular block. Once a transaction is validated by a miner, its block is recorded on the blockchain, and this new addition of the block is distributed across all nodes of the network.

- **Mining**
The transactions in blockchain are verified by blockchain miners in a decentralized manner. During this process of mining, miners are rewarded for authenticating the transaction. Different participants of blockchain compete with each other on a unified authentication. The one who first finds the new block gets rewarded, and this new block is added to the blockchain. Miners attempt to add the blocks on the longest blockchain considering this likely to be the main chain and getting the higher rewards. It is also possible that two miners mine the two blocks at the same time, blockchain will be accidentally forked off until one of the chains becomes longer. Another shorter chain will be abandoned, and blocks in that chain will be called orphaned blocks.

- **Proof of Work**
In Proof of Work, nodes in the blockchain network mine and add the block in the blockchain by solving a difficult mathematical problem. It allows all the nodes in a decentralized blockchain network to reach a consensus on the transactions happening in the network. This makes the network difficult to tamper with and prevents the intruder from faking the transaction and double-spend the cryptocurrency.

- **Confirmation**

  The transaction has to be approved before it is added to the block in the blockchain. This decision is taken by consensus of the participant of blockchain. The transaction is valid if the majority of the nodes agree to that transaction. Initially, when the transaction happens, it is broadcast to blockchain network. Further, miners verify the transaction and add the transaction block to the blockchain. At this point of time, block has zero confirmations. With each block addition in the blockchain afterwards, confirmation of the transaction increases. The higher the number of confirmations for one transaction, the harder it becomes for the intruder to tamper with.

6.4 Blockchain Enabled Electricity Supply Chain

The smart grid system is facing the challenge of transparent distribution of the high volume of renewable energy to balance the supply and demand of energy. Utilizing the existing smart grid systems, an efficient supply chain system is needed to determine how to deliver the electricity transparently and safely thus making the smart grid more efficient. The blockchain-based solution in the electricity supply chain is emerging as a promising solution that helps to share electricity-related information among stakeholders such as producers, microgrids, prosumers, and consumers(Industrial, commercials, and households) securely and efficiently. Table 6.1 outlines the difficulties faced in conventional EP-SC and how these can be settled assuming blockchain is incorporated in the smartgrid supply chain.

To reflect this interaction between producer (power plant) and distributor (microgrid), a smart contract in solidity is designed to represent producer, distributors, consumers, and prosumers. A request is made by microgrid for electricity which is verified to be genuine by proof of work consensus algorithm. Further, the transaction to distribute the electricity gets validated by other participants of the blockchain network, and the electricity distriution record is updated (Refer Fig. 6.4).

After this smart contract is compiled and deployed on the local blockchain using Ganache and Truffle. Ganache is used to deploy and test decentralized applications (dApp) on a local blockchain. Truffle is a framework for testing the smart contract using Ethereum Virtual Machine. With the help of a dApp which is developed using Truffle and Ganache, distributors, consumers, or prosumers initiate a transaction using Metamask crypto-wallet. Metamask is a cryptocurrency wallet that allows users to manage different kinds of

Table 6.1   Traditional EP-SC and Blockchain enabled EP-SC

Features	Traditional Electricity Supply Chain	Blockchain-enabled Electricity Supply Chain
Traceability	To trace the status of electricity supply requests made by users to distributors, different stakeholders send the details to each other, which may be inconsistent. Suppose there is any sudden change in lead time, price, and electricity shortage. In this case, there can be a delay in providing this information in real-time, which demands an effective track-and-trace solution.	With the blockchain integration, the track-and-trace arrangement allows for confirming the recently associated partner or disseminating any change happening in the supply chain to all the stakeholders in real-time. It also helps to avoid human intercession and postponements. The need to rely on a centralized track and trace solution is eliminated in blockchain-enabled EP-SC.
Trust	Maintaining peer-to-peer interactions with trust among all stakeholders is a challenge in the traditional supply chain.	Blockchain enables effective governance by enhancing communication and maintaining trust among various stakeholders.
Scalability	Scaling the conventional supply chain demands various adjustments to maintain the synchronization of information about the status of electricity records.	The DLT based solution helps to improve scalability by sharing consistent information about the electricity records in real-time.
Cost	Human error in capturing the electricity data and maintaining the records can increase the cost of SCM. Smart grids are hoping to diminish EP-SC expenses by eliminating the possibility of human errors. The presence of different intermediaries also increases the cost of traditional EP-SC.	The EP-SC blockchain-based solution helps capture the electricity data consistently, thus enhancing traceability and decreasing the overall electricity supply cost. Since blockchain brings transparency, trust, and security in EP-SCM, the cost automatically gets reduced. Further, eliminating intermediaries minimizes the cost and risk of fraud in EP-SC.

accounts to interact with the Ethereum blockchain. All the transactions made through this wallet are cryptographically signed thus making it secure. Further, the transaction gets validated by miners, and a block representing that transaction is added to the blockchain supply chain. Smart contracts

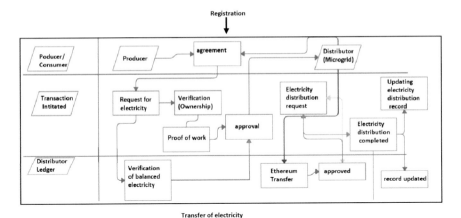

Figure 6.4   Flowchart of proposed blockchain based solution

will automatically trigger the blockchain supply chain when the conditions for distributing the electricity between various stakeholders are met. With the help of blockchain, real-time visibility is introduced into supply chain operations. If any prosumer, consumer or distributor is on-boarded in the supply chain, a blockchain-based supply chain solution records the new stakeholder details and makes it immutable that different participants can trust (Refer Fig. 6.5).

Now with the help of blockchain, all stakeholders will get higher visibility across all supply chain activities. This helps them trace the redundant

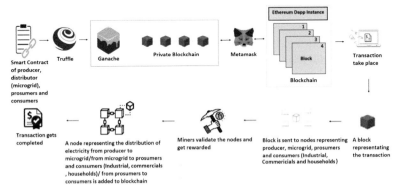

Figure 6.5   Components and technologies used in the proposed blockchain enabled supply chain for smart grid

information in the system, thus making the supply chain system robust and reliable. The Electricity Supply Chain integrated with blockchain will help to create trust between all the stakeholders (producers, distributors, prosumers, residents, commercial users, and industrial users). Users can choose genuine and notable producers and distributors for electricity transfer. Blockchain-based solutions for smart grids will help to create an efficient electricity supply chain focusing on the transparency and security of electricity records thus making the smart grid more efficient. Moreover, introducing blockchain in the electricity supply chain will not only reduces the role of intermediaries but also save the fee of centralized authority by providing a distributed, decentralized, and trustworthy solution for smart grids.

6.5 Conclusion

In this paper, the authors had reviewed the technical bottlenecks in electricity supply chain and discussed how blockchain can be a possible solution to make supply chain efficient and robust. This technology provides transparency in supply chain that allows producers(power plants), distributors(microgrids) and prosumers to obtain the information about the energy consumers(industrial, commercials and households) are demanding. In this chapter, a blockchain-based solution in power supply chain is proposed as a promising solution for securely and efficiently sharing electricity-related information among various stakeholders such as producers, microgrids, prosumers, and consumers. The proposed system allows them to track down any redundant data, making the supply chain system more robust and trustworthy. Further the adoption of blockchain reduces the role of intermediaries in supply chain by creating a secure, decentralised and trustworthy system. The future work will focus on developing a scalable blockchain-based solution for EP-SCM with a modular architecture while preserving privacy.

References

[1] V. C. Gungor, D. Sahin, and et. al, "A survey on smart grid potential applications and communication requirements," *IEEE Transactions on industrial informatics*, vol. 9, no. 1, pp. 28–42, 2013.

[2] X. Fang, S. Misra, G. Xue, and D. Yang, "The new and improved power grid: A survey," *IEEE communications surveys & tutorials*, vol. 9, no. 1, pp. 944–980, 2012.

[3] G. Dileep, "A survey on smart grid technologies and applications, renewable energy," *Renewable Energy Journal*, vol. 146, no. 7, 2020.

[4] A. A. G. Agung and R. Handayani, "Blockchain for smart grid," *Journal of King Saud University - Computer and Information Sciences*, 2020.

[5] X. D. andYing Qi, B. Chen, B. Shan, and X. Liu, "The integration of blockchain technology and smart grid: Framework and application," *Mathematical Problems in Engineering*, 2021.

[6] T. Alladi, V. Chamola, J. J. P. C. Rodrigues, and S. A. Kozlov, "Blockchain in smart grids: A review on different use cases," *Sensors*, vol. 19, no. 11, 2019.

[7] O. M. Butt, M. Zulqarnain, and T. M. Butt, "Recent advancement in smart grid technology: Future prospects in the electrical power network," *Ain Shams Engineering Journal*, vol. 12, no. 1, pp. 687–695, 2021.

[8] Y. Yoldas, A. Ã-nen, S. M. Muyeen, A. V. Vasilakos, and Ã. Alan, "Enhancing smart grid with microgrids: Challenges and opportunities," *Renew. Sustain. Energy Rev.*, vol. 72, pp. 205–214, 2017.

[9] K. Moslehi and R. Kumar, "A reliability perspective of the smart grid," *IEEE Trans. Smart Grid*, vol. 1, no. 1, pp. 57–64, 2010.

[10] N. Gensollen, V. Gauthier, M. Becker, and M. Marot, "Stability and performance of coalitions of prosumers through diversification in the smart grid," *IEEE Trans. Smart Grid*, vol. 9, no. 2, pp. 963–970, 2018.

[11] T. Strasser, "A review of architectures and concepts for intelligence in future electric energy systems," *IEEE Trans. Ind. Electron.*, vol. 62, no. 4, pp. 2424–2438, 2015.

[12] C.-H. Lo and N. Ansari, "Decentralized controls and communications for autonomous distribution networks in smart grid," *IEEE Trans. Ind. Electron.*, vol. 4, no. 1, pp. 66–77, 2013.

[13] F. Hawlitschek, B. Notheisen, and T. Teubner, "The limits of trust-free systems: A literature review on blockchain technology and trust in the sharing economy," *Electronic Commerce Research and Applications*, vol. 29, no. 7, pp. 50–63, 2018.

[14] B. Bhushan, A. Khamparia, K. M. S. andSudhir Kumar Sharma, M. A. Ahad, and N. C. Debnath, "Blockchain for smart cities: A review of architectures, integration trends and future research directions," *Sustainable Cities and Society*, vol. 61, 2020.

[15] S. Saberi, M. Kouhizadeh, J. Sarkis, and L. Shen, "Blockchain technology and its relationships to sustainable supply chain management," *Sustainable Cities and Society*, vol. 57, no. 7, pp. 2117–2135, 2018.

[16] X.-H. Wang and R.-G. Cong, "Electric power supply chain management addressing climate change," *Procedia Engineering*, vol. 29, no. 1, pp. 749–753, 2012.

[17] S. Saberi, M. Kouhizadeh, J. Sarkis, and L. Shen, "Blockchain technology and its relationships to sustainable supply chain management," *International Journal of Production Research*, vol. 57, no. 7, pp. 2117–2135, 2019.

[18] R. Azzi, R. K. Chamoun, and M. Sokhn, "The power of a blockchain-based supply chain," *Computers & industrial engineering*, vol. 135, pp. 582–592, 2019.

[19] S. Saberi, M. Kouhizadeh, J. Sarkis, and L. Shen, "Blockchain technology and its relationships to sustainable supply chain management," *International Journal of Production Research*, vol. 57, no. 7, pp. 2117–2135, 2019.

[20] J. Lukić, M. Radenković, M. Despotović-Zrakić, A. Labus, and Z. Bogdanović, "Supply chain intelligence for electricity markets: A smart grid perspective," *Information Systems Frontiers*, vol. 19, no. 1, pp. 91–107, 2017.

[21] B. Bhushan, A. Khamparia, K. M. Sagayam, S. K. Sharma, M. A. Ahad, and N. C. Debnath, "Blockchain for smart cities: A review of architectures, integration trends and future research directions," *Sustainable Cities and Society*, vol. 61, p. 102360, 2020.

[22] J. Xie, H. Tang, T. Huang, F. R. Yu, R. Xie, J. Liu, and Y. Liu, "A survey of blockchain technology applied to smart cities: Research issues and challenges," *IEEE Communications Surveys & Tutorials*, vol. 21, no. 3, pp. 2794–2830, 2019.

[23] F. Imbault, M. Swiatek, R. De Beaufort, and R. Plana, "The green blockchain: Managing decentralized energy production and consumption," in *2017 IEEE International Conference on Environment and Electrical Engineering and 2017 IEEE Industrial and Commercial Power Systems Europe (EEEIC/I&CPS Europe)*, pp. 1–5, IEEE, 2017.

[24] S. Jabbar, H. Lloyd, M. Hammoudeh, B. Adebisi, and U. Raza, "Blockchain-enabled supply chain: analysis, challenges, and future directions," *Multimedia Systems*, vol. 27, no. 4, pp. 787–806, 2021.

[25] P. Sylim, F. Liu, A. Marcelo, and P. Fontelo, "Blockchain technology for detecting falsified and substandard drugs in distribution: pharmaceutical supply chain intervention," *JMIR research protocols*, vol. 7, no. 9, p. e10163, 2018.

[26] D. Ongaro, *Consensus: Bridging theory and practice.* Stanford University, 2014.

[27] Y. Huang, J. Wu, and C. Long, "Drugledger: A practical blockchain system for drug traceability and regulation," in *2018 IEEE International Conference on Internet of Things (iThings) and IEEE Green Computing and Communications (GreenCom) and IEEE Cyber, Physical and Social Computing (CPSCom) and IEEE Smart Data (SmartData)*, pp. 1137–1144, IEEE, 2018.

[28] J. Gao, K. O. Asamoah, E. B. Sifah, A. Smahi, Q. Xia, H. Xia, X. Zhang, and G. Dong, "Gridmonitoring: Secured sovereign blockchain based monitoring on smart grid," *IEEE Access*, vol. 6, pp. 9917–9925, 2018.

[29] E. Mengelkamp, B. Notheisen, C. Beer, D. Dauer, and C. Weinhardt, "A blockchain-based smart grid: towards sustainable local energy markets," *Computer Science-Research and Development*, vol. 33, no. 1, pp. 207–214, 2018.

[30] F. Jamil, N. Iqbal, S. Ahmad, D. Kim, *et al.*, "Peer-to-peer energy trading mechanism based on blockchain and machine learning for sustainable electrical power supply in smart grid," *IEEE Access*, vol. 9, pp. 39193–39217, 2021.

[31] W. Wang, H. Huang, L. Zhang, and C. Su, "Secure and efficient mutual authentication protocol for smart grid under blockchain," *Peer-to-Peer Networking and Applications*, vol. 14, no. 5, pp. 2681–2693, 2021.

[32] C. Pop, T. Cioara, M. Antal, I. Anghel, I. Salomie, and M. Bertoncini, "Blockchain based decentralized management of demand response programs in smart energy grids," *Sensors*, vol. 18, no. 1, p. 162, 2018.

[33] K. Khatter and R. Devanjali, "Non-functional requirements for blockchain enabled medical supply chain," *International Journal of System Assurance Engineering and Management*, 2021.

7

Virtual Power Plant

Vishnupriyan Jegadeesan[1], Dhanasekaran Arumugam[2], Christopher Stephen[3,4], Ajay John Paul[5], Jahnvi Rajiv Mishra[1], and Vijay Palanikumarasamy[1]

[1]Center for Energy Research, Department of Electrical and Electronics Engineering, Chennai Institute of Technology, India
[2]Center for Energy Research, Department of Mechanical Engineering, Chennai Institute of Technology, India
[3]Department of Mechanical Engineering, Vel Tech Rangarajan Dr. Sagunthala R & D Institute of Science and Technology, India
[4]TARE Fellow, National Institute of Solar Energy, India through Science and Engineering Research Board (SERB), India
[5]School of Mechanical Engineering, Kyungpook National University, South Korea
E-mail: vishnupriyanj@citchennai.net; dhanasekarana@citchennai.net; drschristopher@veltech.edu.in; ajp@knu.ac.kr; jahnvi.eee2020@citchennai.net; vijaykp.eee2020@citchennai.net

Abstract

For more than a century, centralized power generation has ruled modern energy systems. In many countries, the outlook for the grid and generation components of energy infrastructure has not changed. One reason for this is the lack of importance of new technologies. Due to this, the electric grids are continually facing many challenges in various forms of outdated and inadequate infrastructure, which increases demand and causes network blocking, as well as the inability to respond to challenges on time. In recent years, the transitions to energy systems that are both sustainable and decarbonized, as well as energy market deregulation, have resulted in more dynamic and complex structures. The increasing number of local energy consumers and producers, additionally, the growing penetration of distributed generation (DG) from renewable energy sources, posed new encounters to energy investors. The rapid adoption of distributed energy resources (DERs) necessitates advanced technologies, energy plans, and policies to address the techno-economic problems that arise as a result of this adoption. A virtual power plant (VPP) is an idea that refers to central or distributed, cloud-based platforms that can combine, optimize, and control diverse and heterogeneous DERs to function as traditional dispatchable power plants and to deliver power without the physical plant. It consists of small decentralized power plants, storage devices, and controllable loads; thereby, it helps to increase efficiency, flexibility, and reliability. This chapter discusses the need for virtual power plants, their components, communication system architecture, and the benefits of virtual power plants.

Keywords: Distributed energy resources, energy storage, smart grid, virtual power plant.

7.1 Introduction

Today's world is witnessing an unprecedented increase in energy consumption. Two major factors are driving the increase. The first is that we rely on electricity so heavily that even a single day without it can cause irreversible damage. The second reason is that we are living amid a scientific and technological revolution. As a result of these two factors, we discovered that the old system cannot provide us with the desired output or performance [1].

The grid infrastructure, which is responsible for the generation of electricity, continues to operate in many countries in the same way, with no change in its outlook. One of the primary reasons for this is a lack of focus on innovative technologies. The conventional electric grid remains to face several encounters, including outdated and unfit infrastructure, which leads to increased demand, and network congestion issues caused by the grid's inability to respond to challenges in a timely manner. Imbalance manifests itself in the form of blackouts, which are costly for utility companies due to communication issues as well as rising consumer demand to utilize energy usage in order to make the best financial and usage decisions.

Many countries have recognized the technical challenges of the current situation and are focusing their efforts on developing and incorporating advanced technology. DERs and smart grids are two options for dealing with this problem. These grid advancements may hasten the electrification process and reduce periods, hence enhancing service, lowering costs, and ensuring long-term sustainability.

This will allow us to anticipate grid advancements and leapfrog elements of traditional power systems. As stated by Global Health Observatory (GHO), the world's population has grown rapidly, even though the number of people living in urban areas has not yet exceeded the number of people living in the countryside.

As a result, energy demand is expected to skyrocket, creating a massive bottleneck on fossil fuels and diminishing natural resources. To reduce the impact on the earth, it is consequently critical to use renewable and alternative energy sources. Performance with efficiency, optimal generation, demand side management (DSM), and enhanced grid transmission are the entire key to attaining this aim and reducing the amount of money needed to do so.

The cogeneration of locally distributed electricity and heat is aided by employing smaller generation units. Many governments provide incentives to promote distributed energy through renewables, with the chief goal of improving efficiency with the availability of resources. While in the past, it was intended to serve as many active power users as possible, in the near future the energy supply will rise as a result of the involvement of distributed energy units.

The basic concept behind the construction of renewable distributed energy networks was to make them primarily passive. The primary focus was on power distribution with unidirectional power flow to the consumer, whereas the distribution system will need to be more dynamically controlled in the future with a wholly utilized grid, distributed energy resources, and renewable energy units. From a technological and economic standpoint, virtual power plants (VPPs) should be prioritized.

The VPP will offer significant advantages in the form of a more well-organized distribution and control system and allows more energy to be supplied to the feeders, increasing the energy consumed while also allowing for more efficient energy use and the return of unutilized energy to the grid. Transitioning from a traditional to a distributed network approach, such as VPP, will improve sustainability and efficiency while causing long-term change in the energy industry [2].

7.1.1 Distributed Energy Resource (DER)

Energy is created, transmitted, and consumed in the traditional system. Three distinct systems have been combined to form the energy system. The first stage is referred to as production. Power plants generate electricity on a large scale to be cost-effective while also meeting high end-user demand. The second phase is referred to as transmission. Transmission is typically accomplished via overhead power grids or underground power lines. Consumption is the third and final stage. Despite the emphasis on consumption, electrical engineering and accessories are more concerned with the production, transmission, and distribution [3].

DER is a decentralized system. The current system is inefficient to meet the current growing energy demand. DER is generally regarded as a hybrid system that is decomposed into various modules that make it more compatible and convenient.

Generally, conventional power plants are located away from the city, whereas DER is located near the site itself. In addition, DER can use renewable energy, as opposed to the current system, which uses conventional forms of energy. Figure 7.1 shows the DER of a township that generates the energy needed for it rather than relying on the grid supply. Different DER users have different power requirements. Nowadays, hospitals require high reliability and quality of power for the equipment that is highly sensitive. Similarly, the industrial sector has long production hours, resulting in high energy bills, and thermal processes, allowing them to pursue DER applications like cost-effective energy with combined heat and power (CHP) generation.

Computer data centers necessitate consistent, high-quality, and uninterruptible power. To meet these demands, DER technologies are now available and being developed. DER has several components to form a single system; so it is called a distributed system. The main components of a typical DER are generators, storage, and distribution. Generation refers to the production of electricity. These are smaller in scale and easier to operate, manage, and maintain than traditional sources. In general, any source with a capacity ranging from 1 kW to −10 MW can be accommodated in DER. The sources can be fuel cells, microturbines, photovoltaic, wind turbines, and residue of energy.

The DER requires a temporary storage to store energy that can be used according to the requirement. Distribution is a method that is used to distribute electricity on a small scale. Since the grid in DER operates at low voltage, it does not require the usual stepping-up or stepping-down of voltages. The main advantage of this is that, unlike a macro-grid, it is very simple to control. Some of the advantages of DER are as follows:

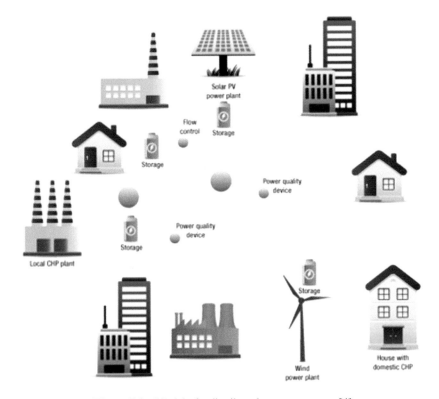

Figure 7.1 Model of a distributed energy resource [4].

- The dependency on the centralized grid is reduced.
- The cost per kWh is low.
- Switching from the power supply and DER is possible.
- Transmission loss is low.
- It uses clean and renewable energy.
- Overhead power lines are eliminated.

Sustainability can be achieved through effective load and energy management.

7.1.2 Smart grid

Smart grid technology includes DER. Contrarily to the traditional systems, where the consumer's only role is to pay the bills, smart grid technology

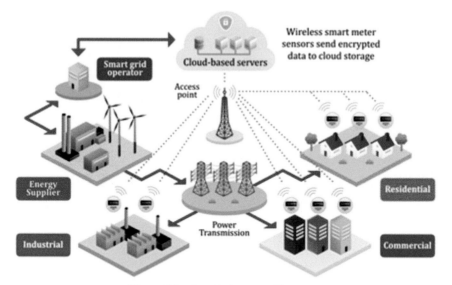

Figure 7.2   A typical smart grid system.

ensures community participation in energy systems. Figure 7.2 shows how a smart grid works [5]. In this system, a consumer should be aware of the way energy is produced, its carbon footprint, and new technologies.

7.2 Virtual Power Plant

A virtual power plant (VPP) is a collection of distributed, medium-scale power generation units, such as solar parks, wind farms, and combined heat and power (CHP), as well as storage systems and flexible power consumers as depicted in Figur 7.3 [6].

VPPs can perform a variety of tasks depending on the market situation. VPP's overarching goal is to connect disseminated energy resources, allowing them to monitor, forecast, optimize, and trade their energy. Variations in renewable energy generation can be compensated for power generation by ramping up and down and controlling power consumption units. It not only stabilizes power grids but also lays the groundwork for the integration of renewable energies into markets.

Due to the variations in generation profile, small/micro power plants cannot provide either balancing facilities or flexibility on exchanges of power, and they do not meet the market's minimum bid size. By aggregating the

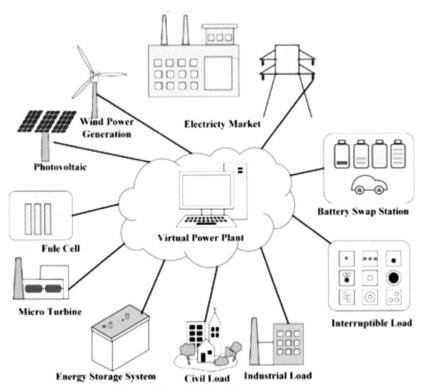

Figure 7.3 A VPP and its components.

power of several units and trading on the same markets, a VPP can offer the same level of service and redundancy as that of large central power plants or industrial customers.

The control system efficiently monitors, coordinates, and controls all assets networked in the VPP. The control data are transmitted over encrypted data connections that are separate from other data traffic. The information required to calculate the best-operating schedules for electricity producers and consumers is stored in the control system.

The central control system, like that of larger conventional power plants, employs an exclusive process to adjust the balancing standby commands from transmission system workers. The bi-directional data transfer between individual plants and VPP enables the transmission of control commands as well as real-time data on capacity utilization in the networked units [7]. The VPP allows for opportunities such as the following.

- Trading: VPP provides energy trading opportunities in competitive electricity markets.
- Network: VPP provides transmission and distribution system operators with system support facilitation services.
- Balancing: Use of multiple markets in real time to balance production and consumption demand.
- Optimizing: The VPP can optimize production and consumption.

VPPs will play a greater role in the future urban energy system by improving smart grid automation and control capabilities by aggregating and virtualizing dispersed DERs.

7.2.1 Components of VPP

A virtual power plant has three main parts [8] that include the following.

- Generation: Generation technology of VPP includes CHP, small power plants and hydro plants, biomass and biogas, wind, and solar energy.
- Energy storage: A unique method of adapting variations in power demand to a constant level of power generation is to use energy storage systems. In cases where the generation is not dispatchable, renewable energy sources can be utilized as a standby. Battery energy storage systems, pumped storage systems, flywheel storage technologies, compressed air storage solutions, supercapacitor storage systems, super conductor magnetic storage devices, and hydrogen in combination with fuel cells can all be taken into consideration for VPP integration.
- Information and communication technology (ICT): The primary requirements for VPP are information and communication technologies and infrastructure. Supervisory control and data acquisition (SCADA), energy management systems (EMS), and distribution dispatching centers (DDC) can all benefit from media technologies.

7.2.2 Classification of the VPP

It can be grouped into two distinct entities [9], which are as follows:

- Technical VPP (TVPP): The TVPP is responsible for managing DGs, controllable loads, and energy storage units, as well as the amendable flow of energy among VPP alliances and auxiliary facilities [10]. It receives details from CVPP regarding agreements on both sides between DGs and consuming units. The information must cover (a) supply and demand forecasting, (b) DG unit allocation and consumption, (c) energy storage division location and capacity, and (d) the largest possible DG unit size [11]. Transmission system operators (TSO) are in charge of bulk power transmission on the central high-voltage electric systems, and it assures the proper operation of the power system while taking into consideration the physical limitations and providing support facilitation services [12]. TVPP functions include determining fault location, providing maintenance facilitation services, monitoring assets, balancing services, managing local networks and implementing ancillary services, providing discernibility to DER units in energy markets, and ensuring that the power scheme operates optimally and safely [13].

• Community-based Virtual Power Plant (CVPP): The amount of energy and the pricing at which CVPP can supply the electrical businesses are of concern to CVPP. It is based on agreements made informally between the generator and consuming units. Small-scale energy producers could not hold on their own in the energy markets. Therefore, CVPP enables them to take part in the energy markets [14]. Power from distributed energy resources (DERs) is sold in the electricity markets, and commerce trading in the commercial power market is one of the capabilities of CVPP. Moreover, CVPP plays a significant role in projecting VPP power output and consumption, preparing and submitting DER bids to the electrical sector, optimizing daily production schedules, balancing trade portfolios, and displaying and involving DER units [15].

7.3 VPP System Architecture

The VPP is a self-contained entity that functions independently under its own rules and constraints to provide adaptability and try to balance services toward the TSO as well as the distribution system operator (DSO). It receives orders from the DSO or TSO control center via an upstream connection. TVPPs are designed to make system administration tasks available to DERs. A TVPP has an impact on the grid; thus, it interacts with the DSO's SCADA and EMS to obtain functional features and data such as power, voltage level, network state, and diagnostics.

One or more CVPPs provide DER data in addition to market data (operating parameters, marginal costs, and metering). A CVPP is only determined by market data and does not consider how DER activation may affect the grid state. The TSO or DSO consolidated VPP accounts containing DER data are routinely sent upstream via VPP. To exchange and transmit essential market-related data, such as bids to the retailer that offers services, the VPP must communicate with a merchant who runs market network applications. For the electricity market, the VPP can get pricing data that can be applied to the utilization of internal capabilities.

Smart grid communication solutions use web services to transport information over TCP/IP architecture. This architecture is enabled by VPP communication. New application-specific standards, such as Open ADR 2.0, which are primarily required for the development of automatic request response programs, DER interface, and VPP implementation, had to be developed to meet the operational requirements of the VPP. In addition to web-based protocols, VPP interchange with other power systems must be enabled. VPP installation also needs to take into account the IEC 61850 interface family, which governs data exchange protocols for a power control system.

Figure 7.4 shows the generalized structure of a VPP. Basis calculation, refinement, and control modules function and make choices within the VPP core. The VPP core makes decisions based on information stored in the database, such as computing baseline values, selecting DERs from the existing portfolio based on predetermined optimization parameters, affordability, sensitivity, and kind, and regulating activation signal delivery. The monitoring module collects the necessary operational data for each reported recipient regularly. Reporting modules can be supplied to SCADA, EMS, or other advanced external entities and are fully programmable.

Weather forecasts and DER product forecasts are two examples of extra functions that contribute to the VPP's overall operation and decision-making.

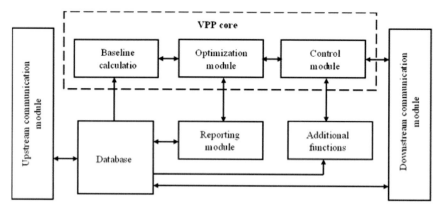

Figure 7.4   Architecture of typical VPP.

Retailers and other third-party services, TSOs/DSOs, and DERs all communicate uniformly via the communication modules [16].

Grid operators can employ virtual power plants to provide ancillary services such as frequency regulation, load following, and operational reserve,
but also aid in grid stability management. The principal goal of these services
is to preserve the electrical supply and demand equilibrium. Because of
the changing levels of consumer consumption, supplementary power plants
must respond to grid operator signals in seconds to minutes to increase or
decrease demand. Future carbon-free electrical networks with substantial
solar and wind power generation must rely on other kinds of controllable
power generation or consumption because ancillary services are typically
provided by controllable fossil fuel generators [17].

To keep the grid operating smoothly, VPPs offer TSOs and DSOs extra
services including intermittent power generation and demand-side management for load−frequency regulation. A capacity reaction mechanism gets
going when a TSO or DSO sends a predefined signal to a VPP [18]. Regardless of the type of action used for load−frequency management, the activation
can be either positive or negative. As seen in Figure 7.5, when there is a grid
outage, either extra generation units or loads must cut usage.

To maintain the correct set-point during negative activation, some DERs
must reduce generation or boost consumption, as shown in Figure 7.6 (the
dashed line).

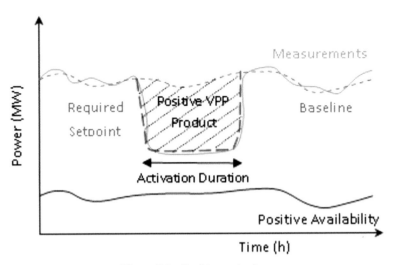

Figure 7.5 Positive activation.

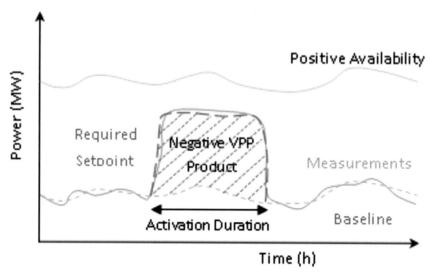

Figure 7.6 Negative activation.

The measurement line shows the DER's electricity generation or consumption (load) (generator). Available capacity is defined as the difference between availability and the measurement line. The capacity is calculated by the VPP during the activation time as the discrepancy between the baseline (dotted line) and measurement's line, and this calculation is shared with the TSO along with other data (or DSO).

7.3.1 Communication system architecture

Figure 7.7 illustrates a hierarchical structure and setting that can be used to describe the communication system. It consists of field area networks (FANs), wide area networks (WANs), neighbor area networks (NANs), and home area networks (HANs). Based on their coverage area and smart grid applications, communication networks are categorized with different quality of service (QoS) needs.

Fiber optics, DSL, and cellular technologies are few examples of communication technologies that can be used to provide a cost-effective solution (2G, 3G, and 4G). TCP is used as a dependable transport protocol in the IP and Ethernet-based VPP communication system. While a virtual private network (VPN) is utilized to meet QoS and security concerns, IP and Ethernet are both stochastic [19].

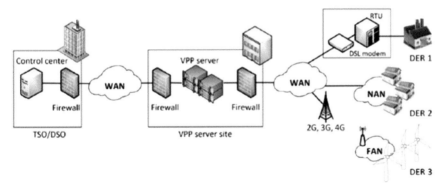

Figure 7.7 Communication system architecture of VPP.

7.3.2 Communication requirements

Units providing supplementary services must satisfy technical and communication requirements defined in the ENTSO-E operational handbook, regardless of traditional power plants or systems adapted in VPP.

The standards specify predefined time of activation (TA) and time of cycle (TC) parameters to achieve the control of load and frequency variation. The full activation times (FATs) and transmission allocations are the intervals of time needed to transmit a signal from either TSO or DSO to VPP and DERs in the downstream direction. The time it takes to gather measurement data and transmit it to the TSO or DSO is referred to as the cycle time, which is abbreviated as TC [20].

Load−frequency control allows TSOs to keep the power system frequency stability in the region synchronous. In a three-step procedure, frequency containment reserve (FCR) is activated, followed by frequency restoration reserve (FRR), manual frequency restoration reserve (mFRR), and automatic frequency restoration reserve (aFRR) action. The ramp-up for the producer/consumer is distinct. Figure 7.8 depicts the ramp-up characteristics.

The timescale represents the approximate time required for a single unit to fully activate. The values on the x-axis show the lowest level that can be curtailed (0%) and the flexible capacity (100%) at its higher level. The choice of a DER for a particular ancillary service is influenced by the operational costs of DERs as well as their technical characteristics.

The ramp-up time is the period when DER production (or consumption) reaches 100% of the set-point value requested by the VPP. The physical and technical characteristics of the DER system have an impact on ramp-up

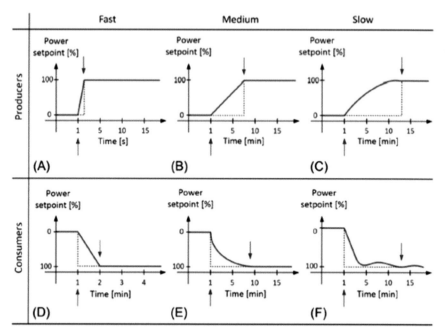

Figure 7.8 Ramp-up characteristics.(1) Producer's group: (a) batteries, (b) industrial steam or gas turbines with a hot start, (c) hydropower run-off river plant, and (d) CHP. (2) Consumer's group: (e) industrial loads like cement mills and (f) steel mills.

times. Ramp-up time affects overall VPP activation latency, which affects how quickly a load−frequency control action reaches the target value TA. Because aFRR and FCR services have significantly higher communication and control requirements, DER units must be carefully taken into account during VPP operation.

7.3.3 Communication technologies

VPP is built around a stable, efficient, and secure communication infrastructure to ensure two-way connectivity and data transfer between various units in the downstream and upstream directions. There are a variety of technologies available to implement the necessary communication facilities for upstream and downstream directions for the physical and link layers. Fiber optics is one of the most commonly used technologies for networks (WAN) [21]. When used for demanding load−frequency control services like aFRR, VPP performance may be impacted by latency and packet loss in cellular connections [22]. Monitoring the important QoS variables is crucial for conducting assessments of the effects of communication abnormalities on VPP performance (latency, packet loss, number of retransmissions, etc.) [23]. Many wireline and wireless technologies are currently used in commercial and research VPP deployments.

7.3.4 Energy management system (EMS)

Energy management is the heart of any VPP. It establishes a long-term and reliable energy data management system. It also collects, stores, and analyzes data from various monitoring devices that are located remotely and determine the best operation plans for both electricity user and producers.

The VPP uses a smart EMS system to coordinate, communicate, and control all internal and external activities like VPP component coordination, DG unit management, as well as storing and consumption management. The EMS is forecasting the output power of the DER units, as well as the controllable loads and power flow. The following are the objectives of VPP [24]:

- lowers the energy production cost;
- improved energy quality;
- reducing the power losses in distribution as well as in transmission systems;
- lessens the greenhouse effect and increases the energy profit margins.

7.4 Challenges to the Implementation of the VPP

VPP faces several implementation challenges. These are divided into (i) technical, (ii) environmental, (iii) regulatory, and (iv) commercial [25].

7.4.1 Technical challenges

Voltage drop, unexpected outages, system capacity, and generation and distribution network operation are all technical issues. Transmission and distribution network operators must deal with the power network system properly to achieve the best results. A more reliable and stable transmission network will emerge by enhancing power quality/profile, lowering transmission losses, and boosting distribution network capabilities, allowing active users to receive electricity without interruption [26].

7.4.2 Regulatory challenges

Because there are few obvious guidelines, it is essential to advance long-term plans and principles to back small-scale contributors [27].

7.4.3 Environmental

The VPP's environmental challenges are related to dispatchable technologies. When integrated, such technologies powered by fossil fuels will impact the environment. These are related to the use of fossil fuel technologies, which are affecting the environment and also have a significant greenhouse effect. Hence, it is critical to carefully evaluate each operating technology implementation.

7.4.4 Commercial challenges

Commercial concerns are worsened by market dynamics, market mechanisms, and market values. New commercial management techniques that reduce network maintenance and operation expenses are used to manage and support small-scale power companies in the distribution network. Incentive techniques can be used to incentivize renewable DG producers. Today's market architecture and pricing models are based on a prolonged procedure. To encourage maximum involvement, market entry tactics can be made simple, easy, and affordable [28].

7.5 Planning of the VPP

By preventing unexpected and unplanned power system failures, network failures, and losses, effective planning saves operational resources. The best operational planning and design of the VPP is the process of setting objectives, standards, and methods for providing power to active users at a minimal price [29, 30]. The purpose of VPP planning is to establish and analyze the plan's technical and economic viability to determine the most cost-effective VPP system with the most benefits [31]–[33].

Technical elements and the maximization of commercial objectives are the primary determinants of optimal VPP planning. Equipment capacities, line loading, wastage, and distribution system operation management, as well as power balancing, voltage regulation, and asset monitoring are all technical components. Lowering the system's total operating cost, boosting system stability and dependability, and reducing transmission losses are all commercial goals [34].

7.5.1 Operation of the VPP

The implementation step entails making the VPP's most affordable plan a reality. The VPP's operational success is determined by several criteria, including consistency, reliability, power security mechanism, and cost competitiveness. The VPP's improved and customized aggregation technique for diverse economic objectives in the power market [35] encompasses the spot electricity market, intraday energy market, and energy market equilibrium via trading links.

Energy exchange, supply and demand management, and system network support assistance services for reducing operating costs and improving profit are all benefits of the VPP system for the power sector [36]. The VPP can be used in the power market to handle a large number of generators while simultaneously delivering real-time data on power production as well as energy optimization.

The EMS can predict data and modify projected information in response to market demand [37]. Using the VPP, wind and solar power may be combined into a single huge power plant, offering power reserves with the common goal of lowering costs and increasing profits.

7.5.2 Advantages of the VPP

The potential advantages of VPP from the perspectives of various entities are as follows [38, 39].

I. Policy makers

- Reduction in global warming.
- Provides additional choices to consumers.
- New opportunities for business and employment.
- Enhances the usage of DER units.
- Creates opportunities for small-scale energy producers.

II. Suppliers and aggregators

- Reduces economic risk.
- New offers.
- Lowers the cost involved in the distribution grid.
- Balancing the deviation in supply and demand forecasting of energy.

- Enables energy efficiency by minimizing the loss in the transmission network.

III. Energy consumers

- Increase in customer satisfaction.
- Continuity of energy supply to consumers during outages.
- Customers will benefit from resiliency services, as well as a reduction in greenhouse gas emissions.
- Vehicle integration into power systems.

IV. DSO and TSO

- Reduces system operators' ancillary services.
- Effective synchronization between DSOs and TSOs, as well as a reduction in grid investments.
- Reduction in the complexity and inflexibility of distributed generation.
- DER unit visibility in distribution or transmission network operation [40].

V. DER owner

- Reduces lowering the cost of entering and operating in the energy market.
- Visualizations of individual DER units in the energy market in real time.
- Enables opportunity for small-scale participants in the energy market.
- Small producers' financial risk is reduced through aggregation.
- Increase in the value of producers' (small) assets [41].

7.6 Artificial Intelligence (AI) and the VPP

A VPP is a cloud-based virtual system that brings together disparate energy resources in one location. The core control system, which is assisted by artificial intelligence (AI) and the Internet of Things (IoT), is the brain of the VPP. These technologies have the potential to create a self-driving grid that can accommodate billions of endpoint users across utility networks and mini-grids. AI is a collection of tools, mathematical models, and algorithms capable of extracting insights from big datasets, identifying patterns, and predicting the probabilities of probable outcomes in complicated, multivariate circumstances.

The most potential AI applications for the energy transition are divided into four categories: renewable energy generation and demand forecasting, grid operation and optimization, energy demand management, and materials discovery and innovation. [42]

Renewable energy resources (RESs) are on the rise, and their fast development in recent years has posed significant issues for power system operators. Energy systems are being compelled to undergo significant transformations to support this new energy generation mix. The majority of renewable energy sources are characterized by fluctuation and intermittency, making power production prediction challenging. These characteristics make power system operation and management more difficult since flexibility is essential for proper operation.

Many countries have set mass deployment goals for cutting-edge metering infrastructure. In the United Kingdom, the gas and electricity markets have set a goal of installing 53 million smart meters for electricity and gas. IoT and AMI generate massive amounts of data, which requires automated data analysis. Furthermore, the transition to increasingly prominent, decentralized, and intelligent power systems necessitates humans executing demanding tasks.

Artificial intelligence is used to estimate power requests and generation and to improve power system stability and efficiency as well as assist the operating personnel by automating decision-making, scheduling, and controlling a plethora of gadgets [43].

To manage decentralized grids during the global transition to renewable energy, artificial intelligence technology will be required. Decentralized energy sources can use AI software to distribute any surplus electricity to the grid, where it can be directed to where it is required. If demand is low, excess energy will be stored in industrial sites, office buildings, homes, and automobiles that AI can subsequently use when production is low.

Moving from an infrastructure-heavy system to AI allows for prediction and control in seconds, which results in a more resilient and flexible grid in the face of unforeseen disasters. All of this means that utilities, policymakers, and regulatory agencies must begin to consider what role they want to play in the development of decentralized energy resources. Coordination and management will be crucial for the patchwork of distributed energy producers. Utilities can take the lead here, as the pool of customers consuming electricity is reducing as more houses and businesses become energy producers [44].

7.7 Case Studies on VPP

Energy & Meteo Systems, for example, began supplying its VPP and power projections to Protergia, MYTILINEOS' Power & Gas Business Unit, in 2020 as a complete IT solution for trading electricity from renewable energy assets. Protein is the largest privately held energy company in Greece. It manages MYTILINEOS' power plants and renewable energy units, as well as trades and supplies electricity and gas to over 280,000 businesses and households, offering modern and dependable services as well as combined electricity and gas packages. Protein is the owner and operator of 1.5 GW of conventional and renewable energy plants.

Energy and meteorological systems provide precise power estimates for all connected wind and solar power facilities, which are then incorporated into the VPP. This gives Protergia greater control over future energy output and reduces the portfolio's exposure to balancing costs. Power forecasting and VPP expertise have been a constant source of support for the VPP, which has been tailored to Protergia's needs [45].

The mathematical models were developed by the faculty of electrical engineering from the inputs from the VPP management system for predicting the amount of electricity used and produced by PVs to achieve the VPP data processing capabilities [46].

In Poland, research was carried out using a VPP that includes hydro-electric plants and energy storage systems (ESSs). For further research, the problems of power quality (PQ) were chosen. To assess the impact of VPP, the study team focused on the use of a global index in a single-point evaluation as well as an area-based assessment [47]. They used 26 weeks of multipoint, synchronic power quality measurements in four separate locations.

7.8 Conclusion

This chapter provides an overview of VPP, as well as its key concepts and benefits. To meet rising energy demands, VPP will play a significant role in future power networks. If DERs are effectively integrated, several energy domain concerns, such as increased peak demand, can be solved. CVPPs and TVPPs, which require particular information translation and communication with different units, are the two forms of VPPs now available for DER integration. DERs can use the VPP idea to get access and visibility in energy markets, as well as profit from VPP market intelligence.

References

[1] https://medium.com/ieee-mec-sb/distributed-energy-resources-8dee35 bef67b-accessedon05-10-2021;11.41am.

[2] Sampath Kumar Venkata chary, Jagdish Prasad, Ravi Samikannu, Challenges, Opportunities and Profitability in Virtual Power Plant Business Models in Sub Saharan Africa – Botswana, International Journal of Energy Economics and Policy, 2017, 7(4), 48-58.

[3] https://medium.com/ieee-mec-sb/distributed-energy-resources-8dee35 bef67b-Accessedon05-10-2021;11.41am.

[4] https://medium.com/ieee-mec-sb/distributed-energy-resources-8dee35 bef67b-accessedon3-3-2022,11.34am.

[5] https://medium.com/ieee-mec-sb/distributed-energy-resources-8dee35 bef67b-accessedon3-3-2022,11.34am.

[6] Hao Bai, Shihong Miao, Xiaohong Ran and Chang Ye, Optimal dispatch strategy of a virtual power plant containing battery switch stations in a unified electricity market, Energies 2015, 8, 2268-2289; doi:10.3390/en8032268.

[7] Zahid Ullah, Geev Mokryani, Felician Campean Comprehensive review of VPPs planning, operation and scheduling considering the uncertainties related to renewable energy sources, IET Energy Systems Integration, doi: 10.1049/iet-esi.2018.004.

[8] Hao Bai, Shihong Miao, Xiaohong Ran and Chang Ye, Optimal dispatch strategy of a virtual power plant containing battery switch stations in a unified electricity market, Energies 2015, 8, 2268-2289; doi:10.3390/en8032268.

[9] H. Saboori, M. Mohammadi, R. Taghe, Virtual Power Plant (VPP), Definition, Concept, Components and Types, 2011 Asia-Pacific Power and Energy Engineering Conference, IEEE Xplore, DOI: 10.1109/APPEEC.2011.5749026.

[10] Glenn Plancke, Kristof De Vos, Ronnie Belmans, Annelies Delnooz, Virtual power plants: definition, applications and barriers to the implementation in the distribution system, 2015, 12th International Conference on the European Energy Market (EEM), IEEE Xplore, DOI: 10.1109/EEM.2015.7216693.

[11] D. Pudjianto, C. Ramsay, G. Strbac, "Virtual power plant and system integration of distributed energy resources," IET Renew. Power Gener., 2007, 1,(1), pp. 10-16.

[12] G. Plancke, Kristof De Vos, Ronnie Belmans, Annelies Delnooz, "Virtual power plants: definition, applications and barriers to the implementation in the distribution system," 2015 12th Int. Conf. European Energy Market (EEM).

[13] A. Richter, I. Hauer, M. Wolter, "Algorithms for technical integration of virtual power plants into German system operation," Adv. Sci., Technol. Eng. Syst. J., 2018, 3, pp. 135–147.

[14] E. Mashhour, S. M Moghaddas-Tafreshi, "A review on operation of microgrids and virtual power plants in the power markets," Second Int. Conf. Adaptive Science & Technology 2009 ICAST 2009 IEEE, 2009.

[15] A. Y. Abdelaziz, Y. G. Hegazy, Walid El-Khattam, M. M. Othman, "Virtual power plant: the future of power delivery systems,"EWRES - II. European Workshop on Renewable Energy Systems, Antalya, TURKEY, 20-30, 2013.

[16] E. Mashhour, S. M Moghaddas-Tafreshi, "A review on operation of micro grids and virtual power plants in the power markets," Second Int. Conf. Adaptive Science & Technology 2009 ICAST 2009 IEEE, 2009.

[17] Matej Zajc, Mitja Kolenc, Nermin Suljanovic, Virtual power plant communication system architecture, Smart Power Distribution Systems, 2019 Elsevier Inc, https://doi.org/10.1016/B978-0-12-812154-2.00011 -0.

[18] https://en.wikipedia.org/wiki/Virtual_power_plant-accessedon07-11-2 021.

[19] O. Palizban, K. Kauhaniemi, J. M. Guerrero, 2014. Microgrids in active network management-part I: hierarchical control, energy storage, virtual power plants, and market participation. Renew. Sust. Energ. Rev. 1, 1–13. https://doi.org/10.1016/j.rser.2014.01.016.

[20] E. Ancillotti, R. Bruno, M. Conti, 2013. The role of communication systems in smart grids: architectures, technical solutions and research challenges. Comput. Commun. 36, 1665-1697. https://doi.org/10.1016/j.comcom.2013.09.004.

[21] R. H. Khan, J. Y. Khan, 2013. A comprehensive review of the application characteristics and traffic requirements of a smart grid communications network. Comput. Netw. 57, 825–845. https://doi.org/10.1016/j.comnet.2012.11.002.

[22] M. Kolenc, N. Ihle, C. Gutschi, P. Nemcˇek, T. Breitkreuz, K. Godderz, N. Suljanovic, M. Zajc, 2018. Virtual power plant using OpenADR 2. 0b for dynamic charging of automated guided vehicles. Int. J. Electr. Power Energy Syst. 104, 370–382.

[23] D. Della Giustina, P. Ferrari, A. Flammini, S. Rinaldi, E. Sisinni, 2013. Automation of distribution grids with IEC 61850: a first approach using broadband power line communication. IEEE Trans. Instrum. Meas. 62, 2372–2383. https://doi.org/10.1109/TIM.2013.2270922.

[24] B. Jansen, C. Binding, O. Sundstrom, D. Gantenbein, 2010. Architecture and communication of an electric vehicle virtual power plant.2010 First IEEE International Conference Smart Grid Commun. (Smart Grid Comm) pp. 149–154. https://doi.org/10.1109/SMARTGRID.2010.562 2033.

[25] M. Kolenc, N. Suljanović, P. Nemček, M. Zajc, 2016. Monitoring communication QoS parameters of distributed energy resources. IEEE International Energy Conference (ENERGYCON 2016). Leuven, Belgium, https://doi.org/10.1109/ENERGYCON.2016.7513900.

[26] S. Ghavidel, Li Li, Jamshid Aghaei, Tao Yu, Jianguo Zhu, "A review on the virtual power plant: components and operation systems," 2016 IEEE Int. Conf. Power System Technology (POWERCON) IEEE, 2016.

[27] R. H. A. Zubo, Geev Mokryani, Haile-Selassie Rajamani, Jamshid Aghaei, Taher Niknam, Prashant Pillai, "Operation and planning of distribution networks with integration of renewable distributed generators considering uncertainties: a review," Renew. Sustain. Energy Rev., 2017, 72, pp. 1177-1198.

[28] X. Liang, "Emerging power quality challenges due to integration of renewable energy sources," IEEE Trans. Ind. Appl., 2017, 53, (2), pp. 855-866.

[29] Ł.B. Nikonowicz, J. Milewski, "Virtual power plants- general review: structure, application and optimization," J. Power Technol., 2012, 92, (3), pp. 135-149.

[30] A. Zurborg, "Unlocking customer value: the virtual power plant," US Department of Energy, 2010.

[31] Fengji Luo, Zhao Yang Dong, Ke Meng, Jing Qiu, Jiajia Yang, Kit Po Wong, "Short-term operational planning framework for virtual power plants with high renewable penetrations," IET Renew. Power Gener., 2016, 10, (5), pp. 623-633.

[32] G. Mokryani, Yim Fun Hu, Panagiotis Papadopoulos, Taher Niknam, Jamshid Aghaei, "Deterministic approach for active distribution networks planning with high penetration of wind and solar power," Renew. Energy, 2017, 113, pp. 942-951.

[33] G. Mokryani, Yim Fun Hu, Prashant Pillai, Haile-Selassie Rajamani, "Active distribution networks planning with high penetration of wind power," Renew. Energy, 2017, 104, pp. 40-49.

[34] Fred S. Ma, David H. Curticer, "Distribution planning and operations with intermittent power production," IEEE Trans. Power Appar. Syst1982, 8, pp. 2931–2940.

[35] K. Rajesh, S. Kannan, C. Thangaraj, "Least cost generation expansion planning with wind power plant incorporating emission using differential evolution algorithm," Int. J. Electr. Power Energy Syst., 2016, 80, pp. 275–286.

[36] J. Haas, F. Cebulla, K. Cao, W. Nowak, R. Palma-Behnke, C. Rahmann, P. Mancarella, "Challenges and trends of energy storage expansion planning for flexibility provision in low-carbon power systems – a review," Renew. Sustain. Energy Rev., 2017, 80, pp. 603–619.

[37] J. Zapata, J. Vandewalle, W. D'haeseleer, "A comparative study of imbalance reduction strategies for virtual power plant operation," Appl. Therm. Eng., 2014, 71, (2), pp. 847–857.

[38] M. J. Kasaei, M. Gandomkar, J. Nikoukar, "Optimal management of renewable energy sources by virtual power plant," Renew. Energy, 2017, 114, pp. 1180–1188.

[39] Y. Degeilh, G. Gross, "Stochastic simulation of power systems with integrated intermittent renewable resources," Int. J. Electr. Power Energy Syst., 2015, 64, pp. 542–550.

[40] Mahmoud M. Othman, Y. G. Hegazy, Almoataz Y. Abdelaziz, "A review of virtual power plant definitions, components, framework and optimization," Int. Electr. Eng. J., 2015, 6,(9), pp. 2010–2024.

[41] M. Braun, "Virtual power plants in real applications-pilot demonstrations in Spain and England as part of the European project FENIX," ETG-Fachbericht- Int. ETG-Kongress 2009, 2009.

[42] S. M. Nosratabadi, H. Rahmat-Allah, E. Gholipour, "A comprehensive review on microgrid and virtual power plant concepts employed for distributed energy resources scheduling in power systems," Renew. Sustain. Energy Rev., 2017, 67, pp. 341–363.

[43] K. El Bakari, W. L. Kling, "Virtual power plants: an answer to increasing distributed generation," 2010 IEEE PES Innovative Smart Grid Technologies Conf. Europe (ISGT Europe) IEEE, 2010.

[44] https://emobilityindia.com/sector-updates/emobility/virtual-power-plant-is-this-the-future-of- energy/01/2021/#:~:text=A%20VPP%20

is%20a%20cloud, management%20centers%2C%20and%20smart %20meters-accessed on 21-02-2022.

[45] Ioannis Antonopoulos, Valentin Robu, Benoit Couraud, Desen Kirli, Sonam Norbu, Aristides Kiprakis, David Flynn, Sergio Elizondo-Gonzalez, Steve Wattam, "Artificial intelligence and machine learning approaches to energy demand-side response: A systematic review", Renewable and Sustainable Energy Reviews,2020, 130, 109899.

[46] https://www.energymeteo.com/customers/customer_projects/virtual-po wer-plant_protergia_greece.php-accessedon-24-2-2022-9.12am.

[47] Tomasz Popławski, Sebastian Dudzik, Piotr Szelag and Janusz Baran, "A Case Study of a Virtual Power Plant (VPP) as a Data Acquisition Tool for PV Energy Forecasting", Energies., 2021, 14, 6200.

[48] Michal Jasinski, Tomasz Sikorski, Dominika Kaczorowska, Jacek Rezmer, Vishnu Suresh, Zbigniew Leonowicz, Paweł Kostyla, Jarosław Szymanda and Przemysław Janik, "A Case Study on Power Quality in a Virtual Power Plant: Long Term Assessment and Global Index Application", Energies., 2020, 13, 6578.

8

AI Business Model is Emerging Energy Market and Smart Grid

Aanal Raval[1], Agrima Lohia[1], and Urvi Y. Bhatt[2]

[1]Area-Information Systems, IIM-Ahmedabad, India
[2]Department of Computer Science and Engineering,
CHARUSAT-DEPSTAR, India
E-mail: aanal.virat6597@gmail.com; agrima11567@gmail.com;
urviybhatt@gmail.com

Abstract

In recent years, AI has fascinated many different industries with its alluring solutions and problem-solving techniques. Among them, it is becoming more and more admirable in energy markets. Its stretching areas include assessing, investigating, and dominating the data of other users via the grid; its decentralization; tackling a large number of participants; maintaining its calculation balance by inspecting the flood of data; integrating electromobility; determining optimal times for the maintenance work of networks; minimization of cost and loss; lessening disturbances; forecasting by evaluating a vast variety of particulars like weather data or historical data; coordination of information about which virtual power plant consumes what amount of electricity and when; as well as handling the stable and green electricity grid. The primary tasks of these industries include blueprinting, installation, and maintenance to deliver a high-quality and secure energy supply. While executing these tasks, one of the greatest challenges faced is to withstand potential uncertainties, which in turn can affect forecasting, scheduling, operation control, and risk management.

AI techniques can handle the whole ecosystem by providing proper planning, monitoring, maintenance, and support for a decision-making system.

This chapter provides a detailed analysis of different models that solve the above issues for better study.

Keywords: Artificial intelligence, energy markets, grid management, machine learning, smart grids

8.1 Introduction

Before anything else, let us have a look at what AI actually is. AI refers to a system that can make intelligent decisions like humans by perception of data, analyzing, and observing it. These systems are capable of decision making and act intelligently through knowledge representation and reasoning, recognizing hidden patterns, and drawing inferences [1]. Machine learning, deep learning, computer vision, natural language processing, etc., are some of its subfields. For the energy sector, as the utilization of AI is increasing, its proficiency and cost of application are decreasing.

A whirlwind furtherance has been noticed in the use of AI in a vast variety of sectors. As a result, in the coming decades, igneous and innovative use of AI in the energy sector has great potential to increase cognizance, congruence, association, effectiveness, capability, reliability, feasibility, and continuous persistence [1]. So, amidst the explosion of processing power, the availability and sustainment of large amounts of data, and many other challenges, AI is able to perform particular targeted tasks without explicit involvement.

Energy markets and realization of AI in energy market applications:

Revenue growth in markets is predominantly driven by the need to acquire operational efficiency to match the requirements of energy [2]. The obligatory condition for the burgeoning use of AI in this sector is its digitalization. Globally, administrations are encouraging the use of AI in energy markets. Issues regarding assessing, investigating, and dominating the data of other users via the grid, its decentralization, and tackling a large number of participants need to be addressed. Changing energy systems need smarter grids. AI has been progressively used in smart grids, trading of electricity, along with its sector coupling, heat, and transport. Solutions involving AI expedite the rationalized process in the energy sector as well as assist in productive, coherent, and secured analysis and computation of gigantic amounts of data [2]. Perseverance of the power grid by determining and solving malformations and anomalies in its creation, triggering, consumption, and transmission stipulates AI enablement. In industries like virtual power plants, electricity markets, oil and gas industries, along with renewable energy plants, the contrivance of AI solutions elevates efficiency and ability to a luminary level.

Based on applications, AI in the energy market can be divided into fleet and asset management, demand response management, renewable energy management, demand forecasting, precision drilling, and infrastructure management [2]. The highest demand for AI enablement can be seen in renewable energy sources as they share the largest revenue and need for low emissions of carbon compounds, which requires frequent and continuous forecasting. AI can estimate and model decisions on storage in terms of demand forecasting, renewable energy generation, issues of network congestion, prices, and weather-load forecasting [2].

Types of energy markets where AI can be enabled [3]:

1. Decarbonization: With imposition of high taxes on the use of fossils to promote a clean and carbon-free environment by promoting the use of renewable sources.
2. Decentralization: It enables the distribution of electricity for a large number of multi-level consumers, geographically.
3. Digitization: At different levels of power system, it promotes the use of digital machines.

8.2 Literature Survey

The authors in [89] presented a secured P2P (peer-to-peer) system for trading for residential buildings and dwellings, aiming to optimize the energy sharing along with operational efficiency of utility vendors in smart grid. They proposed a prosumer recommender system (PRS) with Q-learning and blockchain concerning 6 G for assisting consumers in decision making about selection of prosumers and generators. The Ethereum blockchain and 6G networks are for solving issues related to security and latency, where a buyer will send the request for energy trading to a selected seller, thereafter trading securely with the help of smart contract based on Ethereum blockchain. For energy data management at low cost, they deployed InterPlanetary File System (IPFS). The authors in [90] developed compute-in-memory (CIM) in the form of arrays to address the issue of memory-wall bottleneck, thus improving the energy efficiency. The authors in [91] analyzed deterministic frequency deviations (DFDs) and its relation with external features using eXplainable AI. They used ML models to surveillance daily cycle of DFDs, thereby explaining their interdependencies. The authors in [92] considered uncertainty of high DER (distributed energy resources) penetration for efficient resource management in DER, addressing issues of overloading and grid congestion. The authors in [93] proposed a deep-learning-based method to discover insulator in aerial images. ResNet trained on ImageNet was deployed as a backbone network, cascaded convolution module was employed to extract insulator features in tuning with multiple receptive fields, and finally for learning, spatial pyramid pooling (SPP) and squeeze and excitation (SE) modules were introduced. Integrating two processors with instruction set RISC-V and ARM, [94] created a full stack heterogeneous platform based on open-source neural network accelerator NVDLA. Then deep neural networks were deployed to examine the adaptability to hardware. The work [95] proposed a data management framework for IoT sensors including collection management, storage, quality redundancy, partial labeling, security, and its transformations.

8.3 AI in Energy Market

As a result of digitalization in the energy sector and huge data present for evaluation, AI comes into play. Requisites of AI arise when these voluminous data need to be analyzed securely and efficiently. The following sections provide an analysis on how AI can be implemented in different sectors.

8.3.1 Smart grid and sector coupling

The central idea of a smart grid or sector coupling of electricity, heat, and transport is the connection of information and energy in two ways [4], unlike in the power grid or electricity grid (legacy grid [5]) where it was unidirectional. In a general sense, it is a combination of information system with traditional power grid. This is a network to deliver the electricity from where it is produced to the final destination at which it is consumed. But to be effective and efficient, it should include intervention of AI and distributed generation. As it is known that distributed generation solves the issue of energy transmission and losses over long distances, reducing dependency on single line, let us move on to AI's involvement in it.

The concern of operating with large congeries of variables, data instances, with high dimension and multiple types of data, that include demand, weather, location and generation information has been increasing for the power-cost of grid. These things in turn can decide for each grid what power will come and how much will it cost [6]. AI can solve some of the issues of smart grid, such as low efficiency for energy, and poor interaction, security, and stability [7]. The growing employment of scaling grids, improvement in electricity markets, and the use of renewable energy power plants introduce unpredictability in operations for the power grid. So below is an analysis of applications of AI to smart grid. Figure 8.1 shows the elements in a smart grid as well as operations of smart grid where AI assistance can be applied.

8.3.1.1 Forecasting the power load (demand and supply prediction)

Due to increased amalgamation of renewable energy sources, uncertainty can be seen in the functioning and performance of smart grids. So load forecasting is the estimation of grid power requirements. Parameters affecting a load are customer segments, schedule, season weather, time, and event [8]. This can be classified into three types: 1) short term, which can estimate load up to hours, 2) mid-term, which can estimate up to weeks, and 3) long-term, which can estimate up to years. The short term is generally applied to control energy flow and long term for planning of power generation [63]. Now, short-term estimation needs scheduling and transfer of energy control and required power per grid. However, mid- and long-term estimation considering historical data of power consumption, weather, temperature, humidity, wind, precipitation, time, population, and demographic customer segmentation can be utilized to chalk out future plants.

1. Short term: The authors in [9] proposed a step-by-step method that included wavelet transform, random vector functional link network, and empirical mode decomposition, where ensemble methods improve accuracy and efficiency. The authors in [12] used deep belief networks with parametric copula models to predict load hourly, comparing the results with extreme learning machines, neural network, and support vector regression. The authors in [14] used hybrid clustering with wavelet networks (WNN) and ANN. But ANN can fail in case of limited number of samples in the early stage of ANN where backpropagation exists with only a single hidden node. The work [10] introduced a pooling-based RNN and increased data variety and volume to solve the issue of overfitting. For time consumption for construction of DNN, the authors in [11] combined DNN with different hidden layers and then eliminated layers with bad performance. But this increased the computing overhead. The authors in [13] used Boltzmann machines with factored conditional restrictions (FCRBM) as training and genetics in wind-driven optimization (GWDO).

2. Mid-term: This type of load forecasting is applied for scheduling, maintenance, coordination of load dispatch, balance demand, and generation [8]. The authors in [15] used DBN to predict the peak power load for a year. The authors in [17] experimented particle swarm optimization, with its validation. In [16], DNN model has been applied with two search algorithms for mid-term load forecasting in power systems. The work [19] used CNN and LSTM. The work [18] deployed SVR with mean absolute percentage error (MAPE) of 3.60. The authors in [20] propose an ensemble learning approach by aggregating exponential smoothing with improved LSTM.

3. Long term: The work [21] surveys comparison among multivariate adaptive regression spline (MARS), ANN, and LR, concluding that MARS performs better. The authors in [25] and [27] showed a method that is a combination of LSTM and RNN to predict long-term dependencies in electric load time series. The authors in [22] used LSTM for hourly predictions. The work [23] considered a hybrid model containing LSTM and GRU with MAPE as loss function with economic features, temperature, population as top-down features (forecasting at group level) and customer net load change, feeder load composition, and adoption growth as bottom-up features (requires gathering of information at group level for forecast) along with historical data. The authors in [24] applied LSTM and GRU with hyperparameter tuning. For the same piece of work, [26] compared ANN, SVM, RNN, *k*-nearest neighbor, Gaussian process regression (GPR), and generalized regression neural network (GRNN). The authors in [28] proposed coalescence of fuzzy logic with ANN. The main reason for using LSTM here is that this model can predict regardless of long intervals and time series delays.

8.3.1.2 Grid stability

It is the potential of the system to either preserve or remain in an equilibrium operation state or to achieve the same after a perturbation. Due to huge dimensions and bulky data collected by phasor measurement units (PMUs) and the WAMS (wide area measurement system), there is a need for application of AI methods [8]. Fragments are as follows:

1. Transient stability: It is the ability of a system to be in synchronization, tolerating perturbation. Traditionally used time domain simulations and direct methods cannot address the issue of huge bulky data; so it needs AI intervention. The work [29] used DT, SVM, and ANN for it, and

[30] improved SVM to address the issue of false or missed alarms. The work [31] used LSTM with RNN to learn temporal data dependencies in input. Apart from this, models like ANN [32], ELM [34, 36], CNN [35, 38] with stacked encoders, and DBN have also been applied.

2. Frequency stability: A system upset upshots imbalance between load and generation. So frequency stability is the potential to preserve steady range of frequency. The authors in [39] combined extreme learning and frequency response model for frequency prediction and control.

3. Small-signal-stability: It can be understood as the capacity of a system to be synchronized for small upsets. The work [40] developed a CNN-based model. The authors in [41] used random forest that is multivariate in nature with regression (MRFR), considering 18 bus tests for the same task.

4. Voltage stability: For this specific task models like ANN [42], SVM [43], decision trees [44], and SVR [45], fuzzy logic and iterated random forests [46] have been applied.

8.3.1.3 Fault assessment and flexible equipment

Equipment with power electronic technology is known as flexible equipment. Parameters affecting it are its own variable structure, strong coupling, and uncertain control variables. So finding and resisting faults in these equipment becomes a more tedious task. Models proposed or used for this task includes ELM with wavelet transformation [47], SVR, ANN, KNN, decision tree, Gaussian process regression based generalized likelihood ratio test [48], and a network with sparse and stacking of autoencoders-along with combining-SVM-and-PCA [51]. The authors in [49] used LSTM and SVM for line trip fault prediction. The authors in [50] proposed a hybrid model of wavelet transform and ELM (extreme machine learning) with double channels to recognize, locate, classify, and indexing of liabilities of transmission lines. The authors in [52] used supervised and unsupervised methods to detect fake data injection.

8.3.1.4 Security

Attacks on smart grids result in failures in operation, synchronization, power supply, and data management and its security. This can even lead to failures in cascades and absolute blackouts. Frequently noticed attacks include false data injections (FDIA) and distributed denial of service (DDOS) [8]. For malicious voltage actions in the low-voltage distribution grid and for intrusion detection, [54] and [55] used ANN. The works [58] and [59] used SVM. The work [61] built a CNN random forest model for detecting electricity theft. KNN is also a preferable model. The work [60] used the isolation forest concept and [57] used RL methods. The authors in [56] used amalgamation of fuzzy clustering, gaming theories, and RL. The authors in [53] proposed a neural network model with the concept of stacking in denoising autoencoder (SDAE) to recognize and categorize four different attacks with an accuracy of 96%.

8.3.1.5 Power generation forecast for renewable energy

For sturdy, systematic, energy-saving, and cost-effective operation, it is necessary to make accurate predictions for the generation of renewable energy. For predictions regarding wind power and photovoltaic power generation, the LSTM model can prove to be a good option.

8.3.1.6 Consumer consumption behavior (communication)

For communication in smart grid, advanced metering infrastructure (AMI) enables a two-way communication between load and utilities [63]. The detection of behavior of timely user consumption, abnormal consumption, and non-invasive load monitoring can be made through AI. For tasks of identifying consumption, clustering can be used to define groups and create segments of customer for personalized services. To analyze power consumption and abnormal behavior, user profiles can be summarized and then identified through feature extraction or classification.

8.3.1.7 Distributed energy resource

DER enables the generation of energy locally where it is being consumed. The start-up of energy data modeling, "Lumidyne Consulting," uses data modeling techniques on distributed energy resources. The company solution SPIDER predicts adoption of distributed energy resources as well as its impact on energy demand to achieve accuracy in planning of energy distribution, along with risk and uncertainty management [3]. Another solution, urban energy, focuses on distributed generation by utilizing building rooftops [3].

8.3.2 Electricity trading

In trading, the role of AI is in improving forecasts and the systematic evolution of large data like historical data and weather data as well as for establishing and maintaining grid stability.

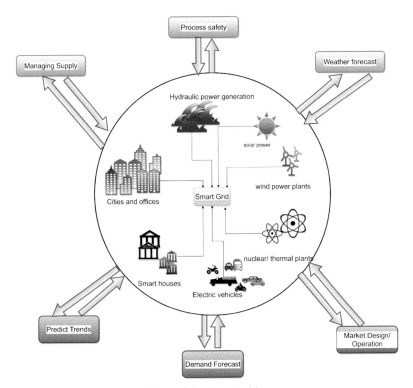

Figure 8.1 Smart grid.

8.3.2.1 Financial market modeling

The authors in [64] used genetic algorithms to create customer segments based on their trading logic. The work [69] discussed the use of ANN in the financial sector for prediction of time series. The work [65] used RCNN for forecasting trading centered on dominant technical indicators and events. The works [66]–[68] used LSTM. To implement high frequency trading and prediction of mid-price movements from input, the authors in [69] discussed the use of RNN to find out the best trading strategy as well as use of NLP for derivative-and-physical-products of forward market, day-ahead market, and intraday with balancing demand-generation.

8.3.2.2 Energy trading forecast

For the task of price forecasting, deployed models are GA [64], behavioral models [70], fuzzy systems [72], neural networks, and hidden Markov model [71].

1. Trade on event fault [69]: The authors in [65] used RNN and CNN to estimate stock price. The work [73] utilized SVM to classify positive and negative stock comments.
2. Trend and technical analysis [69]: The work [74] applied deep learning to forecast pricing, [75] used ANN to estimate the index of the stock market, and [76] deployed a hybrid model of GA and DNN to determine buy and sell points of stock.
3. High frequency: This trade needs constant monitoring of the power system and should be capable of handling large data. This type often includes execution of orders with bulky data at high speed.
4. Price: Generally, application of these includes an analysis of current as well as historic data such as economic environment and carbon policy [69]. So time series and regression are most suitable for these tasks.

8.3.2.3 Energy trading optimization

The work [63] discussed the use of Markov decision process to model trading between consumers and providers as well as RL for optimization. In addition, [63] explored the deployment of Q-learning, deep Q-learning, and fuzzy systems for the same task.

8.3.2.4 Blockchain

It is designed to record transactions in an unambiguous manner using peer-to-peer network and is resistant to transactional data's modification [63]. AI involvement in blockchain includes local trade of energy in smart grid, billing and rewarding based on usage, plug-in vehicles, digital signature, and addressing its leakage and privacy issues [63].

8.3.2.5 Smart meters

Now, in a grid, sending data to a controller at the center of the grid results in more transmission, queuing delays, and bottleneck problem, which affect its real-time operations, privacy concerns, and scalability. So its computations are performed at edge devices, which is known as edge computing. ML can be applied on these edge devices for faster and reliable computation. Google TPU and Pascal GPU deploy ML for the same task [63]. The authors in [77] used deep reinforcement learning for efficient energy scheduling, Q-learning for edge computing, and DNN with cloud for edge server. The work [78] applied LSTM model to detect the nodding movements of a driver for low-cost driving fatigue detection system. The authors in [63] proposed a peer-to-peer energy trading for residential energy management, which can

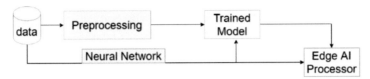

Figure 8.2   Edge computing.

control the operations of heterogeneous DERs and flexible loads that are all DC, in an autonomous, plug-and-play way overcoming physical risks and privacy issues. They used transfer learning for load and price prediction, reinforcement learning for energy trading, data obfuscation for security issues, and edge computing for processing accelerations. Figure 8.2 shows the general operation of edge computing.

8.3.3 Virtual power plant

AI has applications in virtual power plants to determine which plant generates or consumes how much electricity and when. The data required for these tasks are live feed-in data, historical data, data from electricity trading centers, and weather forecasts [62]. Some AI algorithms can trade on its own, which are known as trading algorithms. The authors in [79] used reinforcement-learning-based technique named 'asynchronous advantage actor-critic', along with Neural Networks. The work [82] discussed Q-learning, [80] and [83] used deep Q-learning and explored the usage of LSTM, [84], [86], [87], and [85] used reinforcement and gradient methods, and [88] discussed clustering.

8.3.4 IOE

Internet of Energy applies the concept of distributed control through energy transactions between its users and introduces the idea of smart grid, improving coordination and optimization in energy systems. For example, Energeia – Energy Efficiency Solutions provide monitoring, identification, financing, and implementation of energy-efficient shared-savings business models. Energeia's smart meters gather energy data with the help of gateway devices, which thereafter send the same to an online platform for the task of its analysis [3].

8.3.5 EaaS

Energy as a Service enables electricity selling services, consumption detection, its optimization, and tracking. Being an AI-powered building energy service, IOTA provides a smart building software with energy conservation measures, including lighting, HVAC, demand response, and renewable energy utilization. The main advantage of this system is that it works independent of analytics and communications, lessening the investments [3].

8.3.6 QC

Applications of quantum computing in the energy market aims at introducing and encouraging the use of new solutions to improve efficiency, reducing the use of greenhouse gases. For energy management, JoS Quantum is a cloud-based software solution solving the issues like portfolio analysis, risk analysis, and machine learning (ML) powered enhancements [3]. Another one is QC Ware, which does optimization of energy utilization with ML applications focusing to solve issues like fault diagnosis in energy, its precise prediction, effective demand management, as well as risk analysis of assets [3]. So this platform categorizes the data points for supervised learning and then models are applied on cloud-based simulators [3].

8.3.7 DSM

Demand side management rationalizes the demand to corroborate its harmony with estimated capabilities of the energy system. This rationalization has two parts: demand management for shifting of consumption from peak to off-peak periods, and energy efficiency for monitoring consumption processes. A Portuguese startup, SCUBIC, aims at water networks to refine hydropower management [3]. This platform is capable of forecasting techniques and network simulation and can smartly reduce the operation cost, by actively managing all risks and vulnerabilities.

8.3.8 V2G

Vehicle-to-grid is a system, which enables selling of energy either by returning it to the respective grid or by charging them. Fuergy developed virtual energy networks for each grid, for energy balancing in a sharing system using AI. In the same way, Auto Motive Power provides electric vehicle charging solutions using ampV2G software, which manages energy time and rate and then exports the stored energy back to the local distribution network [3].

8.4 Summary

Papers	Tasks	Techniques used
[9]–[14]	Short-term load forecasting	Ensemble methods, Deep belief networks, WNN, ANN, RNN, DNN, and Boltzmann machines
[15]–[20]	Mid-term load forecasting	Deep belief networks, DNN, SVR, swarm optimization, CNN with LSTM, and ensemble methods with LSTM
[21]–[28]	Long-term load forecasting	Regression-based method, LSTM with RNN, LSTM with GRU, ANN, SVM RNN, K-nearest neighbors, and fuzzy logic with ANN
[29]	Online transient stability assessment	Decision trees, SVM, and ANN
[30]	Real-time transient stability assessment	Improved SVM
[31, 32]	Transient stability assessment system	LSTM with RNN, and ANN
[34, 36]	Transient stability prediction, power system security assessment	ELM
[35, 38]	Transient stability prediction, instability prediction	CNN
[39, 40]	Power system frequency stability assessment	Extreme learning, and CNN
[41]	Security of power grids	Random forest regression
[42]–[46]	Monitoring of voltage stability	ANN, SVM, decision trees, SVR, fuzzy logic, and random forest
[47]	Grid fault detection	ELM
[48]	Fault detection in photovoltaic systems	SVR, ANN, KNN, and Gaussian regression
[49]	Fault prediction in power systems	LSTM and SVM
[50, 51]	Line fault detection	WNN and ELM
[52]	Detect fake data injection	Various supervised and unsupervised methods
[54, 55]	Malicious voltage detection	ANN
[56, 57]	Security for smart grid	Fuzzy clustering and RL

[58, 59]	Intrusion detection for false data injection	SVM
[60]	Data integrity in smart grid	Forests concept
[61]	Theft detection in power grid	CNN, and random forest
[64]	Trading system	GA
[69]	Energy trading	ANN
[65]	Stock market prediction	RCNN
[66]–[68]	Residential load forecasting, and detection of price change indications in financial markets	LSTM
[69]	Estimate stock price	RNN and CNN
[63]	For trading between consumers and providers, and residential energy management	Markov decision process, deep Q-learning, and RL
[77]	Energy scheduling	Deep RL
[78]	Nodding movements of a driver	LSTM
[79]	Driving fatigue detection	RL
[80, 82, 83], [84]–[88]	Trading system	Q-learning, Deep Q-learning, LSTM, RL, and clustering
[89]	Energy trading for residential houses (selection of prosumers or generators)	Q-learning
[90]	To address issues of memory-wall bottleneck	CIM arrays
[91]	Surveillance of daily cycle and features of DFDs	eXplainable AI
[92]	To address issues of overloading and grid congestion	AI techniques
[93]	Discover insulator in aerial image	Deep learning
[94]	To examine the adaptability to hardware	Deep neural networks
[95]	Health state monitoring	IOT data management through ML and big data

8.5 Challenges and Research Gap

With the evolution of smart grids, managing bulky, voluminous, and diverse data (storage, retrieval and determining the needfulness of data) along with online execution of AI/ML techniques has become a whacking challenge. Apart from this, every AI algorithm has its own drawback, which in turn affects applications. With scope and incentive of using renewables, it introduces instability in output due to its divergent and unforeseeable nature.

Major challenges from the AI point of view include the quality of available data, lack of data, tuning of network parameters, issues of compliance, integration, and other technical challenges. Apart from these, updation of outdated infrastructure to meet the needs also cannot be ignored. Determining the right dataset from vast pool available too is inevitable.

With digitization in every field, data becomes a paramount aspect that needs to be concerned. The two-way communication of smart grid is vulnerable to attacks like data injection, data theft, and electricity theft. Although applications have advanced in this field too, whistle-blowers anyway retort to existing security solutions by hook or crook. So this field requires constant upgradation.

Further challenges include regulatory swaps, intricacy in integrating systems, sources, and partnership among various entities in a deregulated environment. As a consequence, it can be observed that some industries tyrannize the market, completely smashing the scope of competition by changing sentiments and attitude of consumers and prosumers.

8.6 Future Directions

Energy market needs to be cost effective, self-healing, adaptive, and automotive. Coalescing with cloud and fog computing increases its robustness, flexibility, scalability, and efficiency. Data tagging is another important aspect that it needs for implementation of machine learning and deep learning algorithms. With advancement in usage of 5G networks, demand side management with automation has become inevitable. Learning consumer consumption and behavior and making decisions on it are an obligation to the smart grid. Till now, many applications of AI have been seen in security and privacy; still, this area needs work on bulky data along with its storage.

8.7 Conclusion

Culminating from these studies, implementation of AI, machine learning techniques for estimation, blockchain, and smart grids (distributed energy resource and Internet of Energy) mutates the business and management models to a new level. Evolved AI techniques are applied to energy market applications with notable results. This chapter discusses possible approaches of AI to energy applications with future directions. Discovering new techniques, spotting state of art opportunities and its implementation with constant remodeling is an unremitting process.

References

[1] IRENA (2019), Innovation landscape brief: Artificial intelligence and big data, International Renewable Energy Agency, Abu Dhabi.

[2] Artificial-Intelligence-in-Energy-Market-By-Product-Offering- (Hardware, AI as a Service, Software), By-Industry-Stream- (Power-Industry, Oil & Gas-Industry), By-Application (Demand-Response, -Fleet-&-Asset, -Renewable-Energy)nd By-Region, -Forecast to-2028 [online] Available at:https://www.emergenresearch.com/industry-report/artificial-intelligence-in-energy-market

[3] Top 10 Energy Industry Trends & Innovations in 2021, StartUs-Insights [online] Available at:https://www.startus-insights.com/innovators-guide/top-10-energy-industry-trends-innovations-in-2021/

[4] I-scoop ,"Smart Grids: electricity networks and the grid in evolution" [online] Available At:https://www.i-scoop.eu/industry-4-0/smart-grids-electrical-grid/

[5] Melike Erol-Kantarci,"Smart Grid? Yes, AI Says: Bring It on!", 2021, IEEE CTN [online] Available At:https://www.comsoc.org/publications/ctn/smart-grid-yes-ai-says-bring-it

[6] Sandra Ponce de Leon, "Role of Smart Grid and AI in the Race to Zero Emissions", Mar 20, [online] Availabl At:https://www.forbes.com/sites/cognitiveworld/2019/03/20/the-role-of-smart-grids-and-ai-in-the-race-to-zero-emissions/h=328b3cb1c8e3

[7] Jian Jiao,"Application and prospect of artificial intelligence in smart grid ", 2020 IOP Conf. Ser.: Earth Environ. Sci. 510 022012, 2020 4th International Workshop on Renewable Energy and Development (IWRED 2020), IOP publish.

[8] Haoran Niu, Olufemi A. Omitaomu, "Artificial Intelligence Techniques in Smart Grid: A Survey", Smart Cities IEEE 2021, 4, 548–568. https://doi.org/10.3390/smartcities4020029.

[9] Qiu, X, Suganthan, P. N, Amaratunga, "Ensemble-incremental learning random-vector functional link network for short-term electric load forecasting." Knowl.-Based Syst. 2018.

[10] Shi-H, Xu-M Li R, "Deep-learning for household load forecasting - A novel pooling deep RNN." IEEE Trans. Smart Grid 2017.

[11] Moon-J, Jung-S, Rew J, Rho-S, Hwang E, "Combination-of short-term load forecasting models-based on a stacking ensemble-approach." Energy Build, 2020.

[12] He Y, Deng J, Li H, "Short term power load-forecasting with deep belief network-and-copula models." In Proceedings of the 2017 9th International-Conference on Intelligent Human Machine Systems and Cybernetics (IHMSC), 26–27 Aust 2017.

[13] Hafeez-G, Alimgeer K. S, Khan I, "Electric load-forecasting based on deep-learning and optimized by heuristic algorithm in smart-grid." Appl. Energy 2020.

[14] Aly, H. H, "A proposed intelligent short term load forecasting hybrid models-of ANN, WNN and KF based on clustering-techniques for smart-grid." -Electr. -Power-Syst. -Res. -2020.

[15] Jiang W, Tang H, Wu L, Huang H, Qi H, "Parallel-processing of probabilistic models based power supply unit mid term load forecasting with apache spark." IEEE Access 2019.

[16] Askari M, Keynia F., "Mid term-electricity load forecasting by a new composite method-based on optimal learning MLP-algorithm." IET Gener. Transm. Distrib. 2019.

[17] Liu Z, Sun X, Wang S, Pan M, Zhang Y, Ji Z., "Midterm-power load forecasting-model based on kernel principal component-analysis and back-propagation neural-network-with-particle swarm optimization." Big Data 2019.

[18] Rai S, De M., "Analysis of classical and machine learning based short term and mid term load forecasting for smart-grid." Int. J. Sustain. Energy 2021, 1–19. doi:10.1080/14786451.2021.1873339.

[19] Gul M. J, Urfa G. M, Paul A, Moon J, Rho S, Hwang E., "Mid term-electricity load-prediction using CNN and Bi-LSTM." J. -Supercomput. 2021, -doi:10.1007/s11227-021-03686-8.-

[20] Dudek-G, Pełka, P-Smyl, S. A, "Hybrid Residual Dilated-LSTM and Exponential Smoothing-Model for Mid term Electric Load-Forecasting." IEEE Trans. -Neural Networks Learn. -Syst. 2021, doi:10.1109/TNNLS.2020.3046629.

[21] Nalcaci G, Özmen A, Weber G. W., "Long term load-forecasting: Models based on-MARS, ANN and LR methods." CEJOR 2019.

[22] Agrawal R. K, Muchahary F, Tripathi M. M., "Long term load forecasting with hourly predictions based on long short term memory-networks." In Proceedings of the 2018 IEEE Texas Power and Energy Conference (TPEC), 8–9 February 20; pp. 1–6.

[23] Dong M, Grumbach L. A, "hybrid distribution feeder long-term load forecasting-method based on sequence prediction." IEEE Trans. Smart Grid 2019.

[24] Kumar S, Hussain L, Banarjee S, Reza, M., "Energy load-forecasting using deep learning approach-LSTM and GRU in spark cluster." In Proceedings of the 2018 Fifth International Conference on Emerging Applications of Information chnology (EAIT), India, 12–13 January 2018.

[25] Bouktif S, Fiaz A, Ouni A, Serhani M. A., "Multi sequence LSTM RNN deep-learning and metaheuristics for electric load forecasting." Energies 2020.

[26] Sangrody H, -Zhou N, Tutun S, Khorramdel B, Motalleb M, Sarailoo-M., "Long term forecasting using machine-learning methods." In Proceedings of the 2018 IEEE Power and Energy Conference at Illinois (PECI), 22–23 February 2018; p. 1–5.

[27] Zheng J, Xu, C, Zhang Z, Li X., "Electric load-forecasting in smart grids using long short term memory based recurrent neural-network." In Proceedings of the 2017 51st Annual Conference on Information Sciences and Systems (CIS, 22–24 March 2017.

[28] Ali D, Yohanna M, Ijasini P. M, Garkida M. B., "Application of fuzzy–Neuro to model weather parameter variability impacts oncelectrical load based on long-term forecasting." Alex. Eng. J. 2018.

[29] Baltas N. G, Mazidi P, Ma J, de Asis Fernandez F, Rodriguez P., "A comparative analysis of decision trees, support vector machines and artificial neural networks for on-line transient stability assessment." In Proceedings of th2018 International Conference on Smart Energy Systems and Technologies (SEST), 10–12 September 2018.

[30] Hu W, Lu Z, Wu S, Zhang W, Dong Y, Yu R, Liu B, "Real time transient-stability assessment in power system based on improved SVM." J. Mod. Power Syst. Clean Energy 2019.

[31] James, J. -Hill, D. J. Lam, A. Y. Gu, J. Li, V. O.,-"Intelligent-time adaptive-transient-stability assessment-system." IEEE Trans. Power Syst. 2017.

[32] Mahdi, M. Genc, V. I. "Artificial-neural-network-based algorithm-for early prediction of-transient stability using wide-area measurements." In-Proceedings of the 2017 5th International Istanbul Smart Grid and Cities Congress anFair (ICSG), 19–21 April 2017.

[33] Mosavi A. B, -Amiri A, -Hosseini, H, "A learning-framework for size and-type-independent transient-stability prediction of power-system using twin convolutional-support-vector-machine." IEEE Access 2018.

[34] Tang Y, Li F, Wang Q, Xu Y, "Hybrid-method for power system-transient-stability-prediction based on two-stage-computing-resources." IET Gener. -Transm. -Distrib. 2017.

[35] Tan B, Yang J, Pan X, Li J, Xie P, Zeng C, "Representational-learning-approach for-power-system transient stability-assessment-based on convolutional-neural-network." J. Eng. 2017.

[36] Liu R, Verbicc G, Xu Y, "A new reliability driven intelligent system for power system dynamic security-assessment." In Proceedings of the 2017 Australasian Universities Power-Engineering Conference (AUPEC) 19–22 November 2017

[37] Wang H, Chen Q, Zhang B, "Transient-stability-assessment-combined-model-framework-based on cost-sensitive method. IET Gener." Transm. Distrib. 2020.

[38] Shi Z, Yao W, Zeng L, Wen J, Fang J, Ai X, Wen J, " Convolutional neural-network based power system-transient stability-assessment and instability mode prediction." Appl. Energy 2020.

[39] Wang Q, Li F, Tang Y, Xu Y, "Integrating model-driven and data driven methods for power system frequency, stability, assessment and control." IEEE Trans. Power Syst. 2019.

[40] Shi Z, Yao W, Zeng L, Wen J, Fang J, Ai X, Wen J., "Convolutional neural network based power system transient, stability, assessment and instability mode prediction." Appl. Energy 2020.

[41] Xiao, Fabus, S. Su, Y. You, S. Zhao, Y Li, H. Zhang, C Liu, Y Yuan, H Zhang, Y et al., "Data Driven Security Assessment of Power-Grids Based on Machine-Learning Approach" Technical-Report; National-Renewable Energy Lab.(NREL): lden, CO, USA, 2020.

[42] Ashraf, S. M, Gupta, Choudhary, D. K, Chakrabarti, S., "Voltage-stability monitoring of power systems using reduced network and artificial-neural-network." Int. J. Electr. Power Energy Syst. 2017.

[43] Mohammadi, H, Khademi, G, Dehghani, M, Simon, D., "Voltage-stability assessment using multi-objective biogeography based-subset selection". Int. J. Electr. Power-Energy Syst. 2018.

[44] Meng X, Zhang P, Xu Y, Xie Hs, "Construction of decision tree based on C4. -5 algorithm for online voltage-stability assessment." Int. J. Electr. Power Energy Syst. 2020.

[45] Amroune-M, Musirin I, Bouktir T, Othman M M, "The amalgamation of SVR-and-ANFIS models with synchronized phasor measurements for online voltage-stability assessment." Energies 2017.

[46] Liu Shi, R Huang, Y Li, X Li, Z Wang, L Mao, D Liu, L Liao, S Zhang, et al. "A data-driven and data based framework for online voltage stability assessment using partial mutual information and iterated random forest." Energies021.

[47] Shafiullah M, Abido M. A, -Al-Hamouz Z, "Wavelet based extreme learning machine for distribution grid fault location." IET Gener. Transm. Distrib. 2017.

[48] Fazai R, Abodayeh -K, Mansouri M, Trabelsi M, Nounou H, Nounou-M, Georghiou, G. E. "Machine learning-based statistical testing hypothesis for fault detection in photovoltaic systems." Energy 2019.

[49] Zhang S, Wang Y, Liu M, Bao Z. "Data based line trip-fault prediction in power systems using LSTM networks and SVM." IEEE Access 2017.

[50] Haq E. U, -Jianjun H, Li K, Ahmad F, Banjerdpongchai D, Zhang T., "Improved performance of detection and classification of 3 phase-transmission line faults based on discrete-wavelet transform and double-channel extreme learning mchine." Electr. -Eng. 2020.

[51] Wang Y, Liu M, Bao Z, Zhang S, "Stacked sparse autoencoder with PCA and SVM for data-based line trip fault diagnosis in power systems". Neural Comput. Appl. 2019.

[52] Ashrafuzzaman M, Das S, Chakhchoukh Y, Shiva S, Sheldon F. T, "Detecting stealthy false data injection attacks in the smart grid using ensemble-based machine learning." Comput. Secur. 2020.

[53] Zhou L, Ouyang X, Ying H, Han L, Cheng Y, Zhang T., "Cyber attack classification in smart-grid via deep-neural-network." In Proceedings of the 2nd International Conference on Computer Science and Application Engineering, 22–24October 2018.

[54] Haghnegahdar L, Wang Y., -A whale, "optimization algorithm trained artificial-neural-network for smart grid cyber intrusion detection." Neural Comput. Appl. 2020.

[55] Kosek A. M., "-anomaly detection for cyber physical security in smart-grids based on an artificial neural-network model." In Proceedings of the 2016 Joint Workshop on Cyber Physical Security and Resilience in Smart-Grids (CPSR-S), 12 April 2016; pp. 1–6.

[56] Wu J, Ota K, Dong M, Li, J, Wang, H, "Big-data analysis-based security situational awareness for smart grid. IEEE Trans." Big-Data 2016.

[57] Ni Z, Paul S, "A-multistage game in smart grid-security: A reinforcement learning solution." IEEE Trans. Neural-Netw. Learn. Syst. 2019.

[58] Zhang Y, Yan J, "Semi Supervised Domain Adversarial Training for Intrusion-Detection against False Data-Injection in the Smart-Grid." In Proceedings of the 2020 International Joint Conference on Neural Networks (IJCNN), 19–24 uly 2020.

[59] Ahmed S. Lee, Y. Hyun, S. H. Koo, "Feature-selection based detection of covert cyber deception assaults in smart grid communications networks using machine learning." IEEE Access 2018.

[60] Ahmed S, Lee Y., Hyun S. H, "Unsupervised machine learning-based detection of covert data integrity assault in smart grid networks utilizing isolation forest." IEEE Trans. Inf. Forensics Secur. 2019.

[61] Li S, Han Y, Yao X, Yingchen S, Wang J, Zhao Q, "Electricity theft detection in power grids with deep learning and random forests." 2019.

[62] "What is Artificial Intelligence in the Energy Industry ?", next kraftwerke, [online] Available at: https://www.next-kraftwerke.com/knowledge/artificial-intelligence

[63] Ning Wang, Shen-Shyang Ho, Jie Li, Chenxi Quie ,"Distributed machine learning for energy trading in the electric distribution system of the future ", Elsevier, 2021.

[64] L. Mendes, P. Godinho and J. Dias, "A Forex trading system based on a genetic algorithm", Journal of Heuristics 2012.

[65] M. R. Vargas, B. S. L. P. de Lima, A. G. Evsukoff, "Deep learning for stock market prediction from financial news articles", 2017 IEEE International Conference on Computational Intelligence and Virtual Environments for MeasurementSystems and Applications (CIVEMSA), 2017.

[66] W. Kong, Z. Y. Dong, D. J. Hill, F. Luo, and Y. Xu, "Short-term residential load forecasting based on resident behaviour learning," IEEE Transactions on Power Systems, vol. 33, pp. 1087-1088, 2018.

[67] Avraam Tsantekidis; Nikolaos Passalis; Anastasios Tefas; Juho Kanniainen; Moncef Gabbouj; Alexandros Iosifidis, Using deep learning to detect price change indications in financial markets, 2017 25th European Signal Processing Conerence (EUSIPCO).

[68] Chenjie Sang, Massimo Di Pierro,"Improving trading technical analysis with TensorFlow Long Short-Term Memory (LSTM) Neural Network" The Journal of Finance and Data Science Volume 5, Issue 1, March 2019, Pages 1-11

[69] Zhibo Ma, Chi Zhang, Chen Qian, "The Development of Machine Learning In Energy Trading", IEEE.

[70] P. R. Kaltwasser, "Uncertainty about fundamentals and herding behavior in the FOREX market", Physica A: Statistical Mechanics and its Applications, 389 (6), pp. 1215-1222, March 2010.

[71] Hassan, M. R., Nath, B.: Stock Market Forecasting Using Hidden Markov Model: a new approach. Proc. of 5th Int. Conf. on intelligent systems design and applications (2005)

[72] O. Badawy and A. Almotwaly, "Combining neural network knowledge in a mobile collaborating multi-agent system", Electrical, Electronic and Computer Engineering, ICEEC '04, pp. 325, 328, 2004, DOI: 10.1109/ICEEC.2004.1374457.

[73] Y. Zhai, A. Hsu, and S. Halgamuge, "Combining News and Technical Indicators in Daily Stock Price Trends Prediction," Lecture Notes in Computer Science, 2007, pp. 1087-1096.

[74] J. Heaton, N. Polson, and J. H. Witte, "Deep learning for finance: deep portfolios," Applied Stochastic Models in Business and Industry, vol. 33, no. 1, pp. 3–12, 2017.

[75] S. T. A. Niaki and S. Hoseinzade, "Forecasting s&p 500 index using artificial neural networks and design of experiments," Journal of Industrial Engineering International, vol. 9, no. 1, p. 1, 2013.

[76] Sezer OB, Ozbayoglu M, Dogdu E. A Deep Neural-Network Based Stock Trading System Based on Evolutionary Optimized Technical Analysis Parameters. Procedia Computer Science. 2017; 114: 473–480.

[77] Yi Liu; Chao Yang; Li Jiang; Shengli Xie; Yan Zhang, "Intelligent Edge Computing for IoT-Based Energy Management in Smart Cities", IEEE, 2019

[78] Chao Yang; Xuyu Wang; Shiwen Mao, "RFID-Based Driving Fatigue Detection", IEEE, 2019.

[79] E. S. Ponomarev, I. V. Oseledets, A. S. Cichock, "Using Reinforcement Learning in the Algorithmic Trading Problem", Journal of Communications Technology and Electronics, 2019.

[80] Y. Deng, Y. Kong, F. Bao, and Q. Dai, "Sparse codinginspired optimal trading system for HFT industry." IEEE Trans. Industrial Inf. 11, 467−475 (2015).

[81] S. Hochreiter and Jü. Schmidhuber, "Long short-term memory," Neural Comput. 9, 1735−1780 (1997).

[82] J. Moody and M. Saffell, "Learning to trade via direct reinforcement," IEEE Trans. Neural Networks 12, 875−889 (2001).

[83] Y. Deng, F. Bao, Y. Kong, Z. Ren, and D. Q. Dai, "Direct reinforcement learning for financial signal representation and trading," IEEE Trans.

Neural Networks & Learn. Syst. 28, 653−664 (2017).

[84] J. Moody, L. Wu, Y. Liao, and M. Saffell, "Performance functions and reinforcement learning for trading systems and portfolios," J. Forecasting 17 (36), 441−470 (1998).

[85] K. L. Xin Du and Jinjian Zhai, "Algorithm trading using Q-learning and recurrent reinforcement," Learn. Positions 1, 1 (2009).

[86] R. J. Williams, "Simple-statistical gradient following algorithms for connectionist-reinforcement learning," Machine Learn. 8, 229−256 (1992).

[87] Thibaut Tht'eate, Damien Ernst, "An Application of Deep-Reinforcement Learning to Algorithmic Trading", arxiv:2004.06627v3, 2020

[88] Simon Kuttruf, "A Machine Learning framework for Algorithmic trading on Energy markets", 2018 Available Online At:https://towardsdatas cience.com/https-medium-com-skuttruf-machine-learning-in-finance -algorithmic-trading-on-energnergy-markets-cb68f7471475

[89] A. Kumari, R. Gupta and S. Tanwar, "PRS-P2P: A Prosumer Recommender System for Secure P2P Energy Trading using Q-Learning Towards 6G," *2021 IEEE International Conference on Communications Workshops (ICC Workshops)*, Montreal, QC, Canada, 2021, pp. 1-6, doi: 10.1109/ICCWorkshops50388.2021.9473888.

[90] Xin Si, Yongliang Zhou, Jun Yang, Meng-Fan Chang, "Challenge and Trend of SRAM Based Computation-in-Memory Circuits for AI Edge Devices", 2021 IEEE 14th International Conference on ASIC (ASICON), 2021

[91] Johannes Kruse, Benjamin Schäfer, Dirk Witthaut, "Exploring-deterministic frequency deviations with explainable AI", IEEE International Conference on Communications, Control, and Computing Technologies for Smart Gids (SmartGridComm), 2021

[92] José Almeida, João Soares, Fernando Lezama, Bruno Canizes, Zita Vale, "Evolutionary Algorithms applied to the Intraday Energy Resource Scheduling in the Context of Multiple-Aggregators", EEE Symposium Series on Computatonal Intelligence (SSCI), 2021

[93] Zheng Huang, Hongxing Wang, Bin Liu, Jie Zhu, Wei Han, Zhaolong Zhang, "Insulator Detection Based on Deep Learning Method in Aerial Images for Power-Line Patrol", 11th International Conference on Power and Energy Systems (ICPES, 2021

[94] Zikang Zhou, Chao Fu, Ruiqi Xie, Jun Han, "A Heterogeneous Full-stack AI Platform for Performance Monitoring and Hardware

specific Optimizations", IEEE 14th-International Symposium on Embedded-Multicore/Many-core Systems-on-Chi(MCSoC), 2021

[95] Wenjin Yu, Yuehua Liu, Tharam Dillon, Wenny Rahayu, Fahed Mostafa, "An Integrated-Framework for Health State Monitoring in a Smart Factory-Employing IoT and Big Data Techniques", IEEE Internet of Things Journal, 2022

9

Artificial Neural Network and Forecasting Major Electricity Markets

Vaibhav Aggarwal[1], Sudhi Sharma[2], and Adesh Doifode[3]

[1]O. P. Jindal Global University, India
[2]Fortune Institute of International Business, India
[3]Symbiosis School of Banking and Finance, Symbiosis International University, India
E-mail: vaibhavapj@gmail.com; sudhisharma1983@gmail.com;
equity.adesh@gmail.com

Abstract

Coal and natural gas are some of the major energy sources, and their price movements significantly influence industries and the economy. Traditionally, econometric models have been used to forecast energy commodity prices. However, coal and natural gas prices are estimated using ARIMA and artificial neural networks (ANN) in this study. Daily prices of coal and natural gas from 1 January 2010 to 11 November 2021 are sourced from Bloomberg terminal. Every country in the world is directly or indirectly dependent on coal for its electricity requirements in the industrial and residential sectors. The countrywide utility of these essential commodities can also impact the reserves in an economy. Thus, this study is novel and contributes to the existing literature with its findings. In addition to the academic contribution, the findings will benefit policymakers in making broader decisions, the management of the companies in their procurement decisions, and investors in general.

Keywords: ARIMA, commodities, forecasting, Natural gas, neural network.

9.1 Introduction

Energy has become the fundamental need of humanity and a necessity for all economies across the globe. Thus, economies are substantially dependent on an uninterrupted supply of different forms of energy for their smooth functioning. Critical commodities of energy sources like coal, crude oil, natural gas, etc., are directly used in generating electricity or fuel [1], while indirectly, commodities like copper, zinc, lead, etc., are used. Kaufmann and Hines [2] found a significant correlation between the prices of these three fundamental energy sources, viz. coal, crude oil, and natural gas, in addition to the past literature [3–5]. Movement in energy prices impacts all industries directly or indirectly [6].

Developed countries are highly industrialized, leading to increased consumption of global energy resources. According to the US Energy Information Administration (EIA) projection in the Annual Energy Outlook 2021 [7], the industrial and electric power sector is likely to propel energy consumption in the US, as shown in Figure 9.1. In the annual industrial energy consumption share, the manufacturing sector has the largest share, with around 77%, followed by the mining and construction sectors with about 12% and 7%, respectively. Petroleum and natural gas consumption will increase in the coming three decades, while a substantial increase in renewable energy sources is foreseen during the same period, exhibited in Figure 9.2.

Energy resources across the globe are gradually depleting due to easy exploitation by mining companies [8]. The use of coal by industries in an economy increases with the increase in crude oil prices [9]. In developing countries, energy demand and consumption are lower than the global average per capita of 1.29 toe but are likely to increase rapidly. They are likely to observe explosive growth in renewable energy generation. India is improving energy security through significant investments in strengthening its renewable capacities and allowing 100% FDI in the power sector. The Indian government aims to enhance natural gas share in the energy mix from 6% in 2020 to around 15% by 2030 [10]. India is progressing well toward achieving targets set by the United Nations for sustainable energy, and India is actively involved internationally in reducing its per capita carbon emissions.

Increasing energy prices result in cost-push inflation in an economy and finally increases wholesale price index (WPI) inflation [11, 12]. Thus, more accurate forecasting models are helpful and vital in reducing spending on a country's import bill and helping the governments take the right decisions using the early signals. Chen *et al.* [13] also established the significance

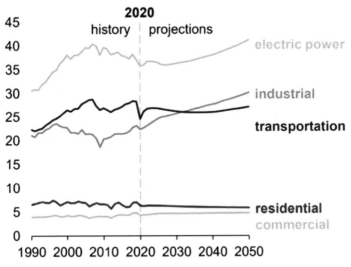

Figure 9.1 US energy consumption by sector [7].

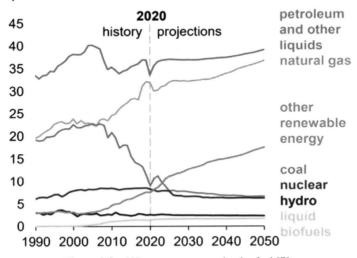

Figure 9.2 US energy consumption by fuel [7].

of exchange rates in addition to other variables in the energy forecasting process while also identifying that exchange rates strongly predict the energy prices of exporting economies. Contradicting previous studies, Sun and De [14] found that the volatility in currency impacts both exporting and importing economies. Most economies get impacted by the movement in prices of these fundamental commodities. Therefore, it becomes imperative and critical to study and forecast energy and raw material price movements to get a timely indication for appropriate economic decisions. Thus, volatility estimation associated with energy prices is essential for improved regulators' decision-making accuracy.

Volatility forecasting of energy commodity prices has attracted industry practitioners and academicians over the past few decades [15]. Additionally, linkages between commodity prices like precious metals and other metals like iron, copper, etc., with energy prices are investigated by researchers [16–21]. Volatility in energy raw material prices and related commodities significantly impacts electricity and fuel prices in unregulated markets [22], and the accuracy of these forecasting models highly depends on these input variables.

Models for forecasting energy prices evolved from traditional statistical models [23–25] to machine learning models [26, 27] and advanced further to hybrid models [28]. Comprehensible and easy-to-use price forecasting techniques must be adapted and effective with a credible outcome as these research works are offered to potential investors in financial markets [29]. Thus, a blend of time series forecasting and machine learning models predicts with more accuracy [25].

The feasibility studies in energy prices are complex for the investors to apprehend, and thus a possibility of lucid predicting models is required where the price estimations are confined [30, 31]. Extensive literature is available on predicting energy and raw material prices using econometric models, while few recent studies have also forecasted using hybrid models [32–36]. Bento *et al.* [37] used artificial neural networks based on Bat algorithm for predicting energy prices for the short term.

Yu *et al.* [38] forecasted crude oil by testing support vector machine methods using artificial neural networks, ARIMA, ARFIMA, and Markov-switching ARFIMA model. Similarly, Čeperić *et al.* [39] predicted natural gas prices using feature selection algorithms for improved estimates by applying neural networks. Cuaresma et al. [40] used univariate autoregressive-moving average models using hourly electricity spot prices to predict electricity prices.

Researchers [41, 42] have also attempted to predict raw material prices, viz. uranium prices and coking coal prices, using ARIMA models. Additionally, He *et al.* [43] and Lasheras *et al.* [44] applied ARIMA models using neural networks in estimating precious metal prices and copper prices, respectively. These past studies have prompted the authors to investigate coal and natural gas price volatility using neural networks.

Different models have evolved in the past literature, applied to forecast a broader range of raw materials linked with energy. Matyjaszek *et al.* [27] applied ARIMA models and neural networks to predict coking coal movements using monthly spot data for 25 years from January 1991. The vector autoregression (VAR) model [45] was used before the honey bees mating optimization algorithm [46], which had better precision in the estimation of energy prices. Further, using machine learning, Herrera *et al.* [26] estimated coal, crude oil, and natural gas prices. Kong *et al.* [47] and Mengdi and Yong [48] developed system dynamics and structural time series modeling to estimate long-term coal and carbon prices. Volatility is an important factor in investment decision-making, and different GARCH models have been used in various forecasting studies [49–52]. At the same time, recent studies have used neural network models in the forecasting of natural gas [53–55].

9.2 Research Methodology

9.2.1 Data profile

The study aims to predict the underlying commodities, i.e., coal and natural gas, using artificial neural networks. Daily closing prices from 1 January 2010 to 12 November 2021 are sourced from the Bloomberg terminal. The returns have been calculated by applying them. The variations in the dates of both asset classes have been handled, and, finally, at this stage, massage data is ready for further forecasting. The R studio has been used for the predictive model.

9.2.2 Methods and models

The study has applied predictive models, i.e., ARIMA and ANNs. However, before applying the models, the study provides the behavior of the asset class closing prices and its returns. The detailed models used in this chapter have been discussed hereunder.

9.2.2.1 Descriptive statistics

The closing prices and returns of both commodities, i.e., coal and natural gas, have been visualized through descriptive statistics in terms of maximum, mean, standard deviation, and skewness.

9.2.2.2 ARIMA model

The traditional method of forecasting time series is an autoregressive integrated moving average (ARIMA), developed by Box−Jenkins. The ARIMA model depends on three components, such as autoregression (AR), integration (I), and moving average (MA). It takes the form of *p,d,q* specification, representing autoregression, integration, and moving average, respectively. AR means the current value depends on its lagged value. "I" means the level of integration/stationarity. The ARIMA model depends on the level of integration or stationarity. The ARIMA model applied can be applied either on level, i.e., I(0), or at the first-order difference, i.e., I (1). But if the time series is stationary at the second-order difference, then it could be considered a non-stationary time series and then for prediction, we are moving toward non-linear models. MA means the error term of the basic equation of both coal and natural gas are:

$$cp(coal)_t = a_0 + a_1 cp(coal)_{t-1} + ut$$

where $cp(coal)_t$ is the closing price of coal at time t, $cp(coal)_{t-1}$ lagged closing price, and Ut is error term.

$$cp(NG)_t = a_0 + a_1 cp(NG)_{t-1} + ut$$

where $cp(NG)_t$ is the closing price of natural gas at time t, $cp(NG)_{t-1}$ is lagged closing price, and Ut is the error term.

ARIMA models can be estimated by following the Box−Jenkins approach. The ARIMA model for coal and natural gas has been formulated in eqn (9.3) and (9.4), as follows:

$$cp(coal)_t = a_0 + a_1 cp(coal)_{t-1} + a_2 cp(coal)_{t-2} + \ldots$$
$$a_n E_{t-n} + a_{n+1} E_{t-2} \ldots \ldots a_{n+1+m} E_{t+n+1+m} + \varepsilon t \ldots \quad (9.1)$$

where $cp(coal)_t$ is the closing price of coal at time t, $cp(coal)_{t-1}$ is the lagged closing price, and E_{t-n} is the lagged error.

$$cp(NG)_t = a_0 + a_1 cp(NG)_{t-1} + a_2 cp(NG)_{t-2} + \ldots$$
$$a_n E_{t-n} + a_{n+1} E_{t-2} \ldots \ldots a_{n+1+m} E_{t-n+1+m} + \varepsilon t \ldots \quad (9.2)$$

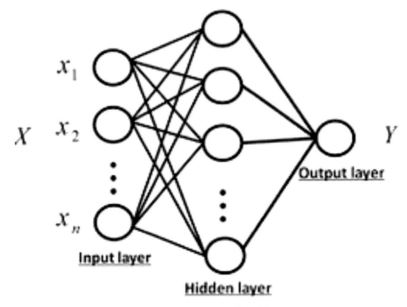

Figure 9.3 Layers of ANNs.

where $cp(NG)_t$ is the closing price of natural gas at time $t, cp(NG)_{t-1} is$ lagged closing price, and E_{t-n} is the lagged error.

9.2.2.3 Artificial neural networks

A robust forecasting model, i.e., artificial neural networks (ANNs) as shown in figure 9.3, is used for predicting time series models. Ljung–Box tests considered ANNs a more accurate model to predict a time series' prices. The superiority of the model is that the model could be applied in both linear and non-linear models. It means that if the time series is stationary at second-order difference or non-stationary, then ARIMA models are superfluous. In such cases, ANNs are applicable and provide better results. The infrastructure of ANNs constituted of three layers named input, hidden, and output layers. Time series prediction based on its own lagged values is considered a shallow neural network. The infrastructure of ANN used in the chapter to predict the closing prices of both coal and natural gas is as follows:

Finally, after application of the models, the study identifies the best-suited model on the basis of mean error (ME), root mean squared error (RMSE), mean absolute error (MAE), mean percentage error (MPE), and mean absolute percentage error (MAPE). The different variants of errors would be the criterion to select the robust model. The lower the values, the higher would be the predictive power of the model.

9.3 Empirical Discussion

9.3.1 Discussion on descriptive statistics

Table 9.1 encapsulated the results of descriptive statistics of absolute closing prices of the underlined energy-driven asset class, coal, and natural gas. Further, the descriptive statistics extended to the returns of underlined energy assets. The number of observations of absolute closing prices is 3096. The average closing prices of coal and natural gas are USD 88.24 and USD 3.23, respectively. However, investors are more interested in the results of the descriptive of returns. The maximum return is given by natural gas (20%) followed by coal (14%). However, the standard deviation is high in natural gas (0.03), followed by coal (0.02). It means the asset class follows the principle of high returns and risk. Another interesting piece of evidence is perceived from skewness. The natural gas skewness is (0.27), whereas coal has negative skewness (-6.54). It means that natural gas has a high probability of getting positive returns, and coal has a probability of getting negative returns. The kurtosis of coal is significantly higher than natural gas. It means higher peakedness perceived in coal prices. It means that there is high volatility in coal prices than natural gas. These inferences drawn from descriptive are very important for the investors looking for investments in underlined assets of the energy sector.

Table 9.1   Results of descriptive statistics.

	Coal	Natural gas	Coal_returns	Natural gas_returns
Nobs	3096	3096	3095	3095
Minimum	48.50	1.48	−0.43	−0.18
Maximum	269.50	6.31	0.14	0.20
Mean	88.24	3.23	0.00	0.00
Median	86.60	3.00	0.00	0.00
Sum	273,183.68	9998.38	0.53	−0.09
Variance	750.62	0.78	0.00	0.00
Std. dev.	27.40	0.88	0.02	0.03
Skewness	1.55	0.52	−6.54	0.27
Kurtosis	5.59	−0.17	211.68	3.88

9.3.2 Discussion on plots

After understanding descriptive statistics according to the investor's perspective, the study understands the behavior of both time series with its plot shown in Figures 9.4 and 9.5. The coal showed volatility in three phases from 2010 to 2016, followed by 2016 to 2020, and afterward suddenly showed a huge spike. Natural gas, as already perceived in descriptive statistics as a comparatively high return energy commodity, the same is captured from the plot. From the technical analyst's perspective, there are three bottoms and four peaks. From the understanding of the plots, the investor will consider natural gas as more attractive than coal because coal from 2010 to 2020 is almost range bound. By plotting the trend line in the plots shown in Figures 9.6 and 9.7, it has been perceived that both the assets are stochastic rather than deterministic. Furthermore, natural gas means is more consistent than coal. After well-understanding the data of both commodities, now, the discussion moves forward to predicting prices from various models.

The study has considered absolute values of closing prices of both coal and natural gas its log values. The reason behind the same is the perceived volatility and then the variations in the closing prices of both commodities. Thus, log values have been considered for both coal and natural gas: Ln(Coal) and Ln(NG) to mitigate these issues.

9.3.3 Discussion on the results of ARIMA predictive models

The study first checks the level of stationarity in the log time series of both coal and natural gas by applying augmented Dickey−Fuller (ADF) test. With the application of the ADF test, the stationarity level has been found to

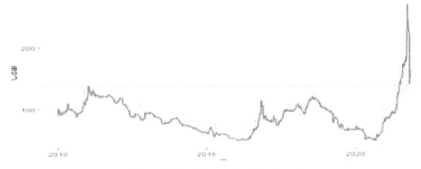

Figure 9.4   Time series plot of coal.

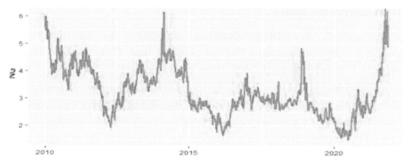

Figure 9.5   Time series plot of natural gas.

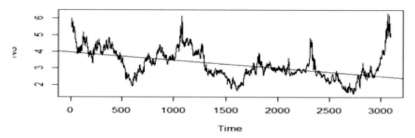

Figure 9.6   Plot of time series and trend of coal.

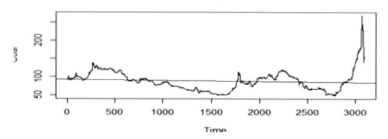

Figure 9.7   Plot of time series and trend of natural gas.

provide the form of d of p,d,q. Second, the study has visualized the ACF and PACF that provide the order of q and p, respectively.

9.3.4 Results of stationarity:augmented Dickey−Fuller (ADF) test

From Figures 9.6 and 9.7, we have found trends in the prices of coal and natural gas. Thus, it is crucial to gauge the stationarity assumptions. Otherwise, the results of ARIMA on non-stationary time series will be

Table 9.2 Results of augmented Dickey–Fuller (ADF) test.

lnCoal
Dickey–Fuller = −1.3962, lag order = 14, *p*-value = 0.834
lnNG
Dickey–Fuller = −2.3312, lag order = 14, *p*-value = 0.4381
d_lnCoal
Dickey–Fuller = −11.96, lag order = 14, *p*-value = 0.01
d_lnNg
Dickey–Fuller = −15.167, lag order = 14, *p*-value = 0.01

spurious. The results of the ADF test are captured in Table 9.2. The results at level show that the log values of closing prices of commodities are not showing stationarity. The coefficients are negative, i.e., −1.39 and −2.33, but *p* values are greater than 0.05, which rejects the null hypothesis of stationarity, and thus the series is non-stationary at the level. It means that there is a unit root or trend; thus, both series are not stationary at level I(0). Now, removing trends from the prices of both commodities is crucial. The presence of the availability of trend means the mean and variance are not constant over time. Then the first-order differencing has been done, and then again, ADF test has been applied on both the series that are stationary. The results shown in Table 9.2 found that the series at first-order differencing are stationary. The coefficients are negative, and the *p*-value is less than 0.05; thus, the null hypothesis is accepted, and both the series are stationary or integrated at I (1). Henceforth, for further prediction, first-order differencing series have been used.

9.3.4.1 Results of ACF and PACF

After determining the level of stationarity, the next step of ARIMA is to get a plot correlogram to identify the lagged values of autoregression (AR) and moving average/error term (MA). The plots of auto-correlation function (ACF) and partial auto-correlation function (PACF) are shown in Figure 9.8. ACF will give the order of MA, and PACF will give the order of AR.

The results of plots of ACFs and PACFs have been shown on both log coal (lnCoal) and log natural gas (lnNG). The results of ACF and PACF provides evidence of current price dependencies on lagged values thus, ARMA specifications are applicable.

9.3.4.2 Results of Auto.ARIMA

Since the study is using R, a very robust function, i.e., auto.ARIMA, is available to detect the form of ARIMA *p,d,q* specifications. Tables 9.3 and 9.4

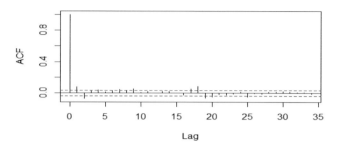

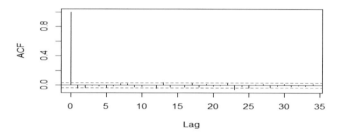

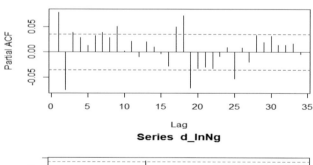

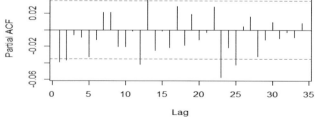

Figure 9.8 Plot correlogram − ACF and PACF.

Table 9.3 Results of Auto.ARIMA of log coal.

Series: lnCoal
ARIMA(3,1,3)
Coefficients:
ar1 ar2 ar3 ma1 ma2 ma3
1.3718 −0.1368 −0.3105 −1.292 −0.0403 0.4291
s.e. 0.1130 0.2014 0.1047 0.107 0.1885 0.0974
sigma∧2 = 0.00024: log likelihood = 8509.39
AIC = −17004.79 AICc = −17004.75 BIC = −16962.53
Training set error measures:
ME RMSE MAE MPE MAPE
Training set 0.00013 0.0154 0.0068 0.00241 0.1540

Table 9.4 Results of Auto.ARIMA of log natural gas.

... Series: lnNG
ARIMA(2,1,2)
Coefficients:
ar1 ar2 ma1 ma2
−0.2101 0.6434 0.1761 −0.6880
s.e. 0.2672 0.2566 0.2519 0.2413
sigma∧2 = 0.0008495: log likelihood = 6552.61
AIC = −13095.21 AICc = −13095.19 BIC = −13065.03
Training set error measures:
ME RMSE MAE MPE MAPEc
Training set −0.00003 0.0291 0.0210 −0.0535
2.0253

show the results of auto.ARIMA. The results show that for log, coal(3,1,3) is the best-suited model. It means the values are integrated at first-order differencing, and current values depend on three lagged values and three lagged error terms. For log natural gas, (2,1,2) is the best-suited model. It infers that the closing prices of log natural gas are integrated at I (1), and further, the current value depends on two lagged values and two lagged error terms. The robustness of the best-suited model is shown in Table 9.5.

9.3.4.3 Forecasts of the model

The forecasts of the model have been presented in Figures 9.9 and 9.10. Figure 9.9 is showing the forecast plot of coal and Figure 9.10 is showing that of natural gas.

Table 9.5   Results of best-suited ARIMA model.

ARIMA(p,d,q)	Coal gas	ARIMA(p,d,q)	Natural gas
ARIMA(2,1,2)	-16981.15	ARIMA(2,1,2)	-13093.7
ARIMA(0,1,0)	-16943.88	ARIMA(0,1,0)	-13083.18
ARIMA(1,1,0)	-16962.53	ARIMA(1,1,0)	-13088.32
ARIMA(0,1,1)	-16964.14	ARIMA(0,1,1)	-13086.19
ARIMA(0,1,0)	-16945.51	ARIMA(0,1,0)	-13085.17
ARIMA(1,1,2)	-16976.47	ARIMA(1,1,2)	-13093.16
ARIMA(2,1,1)	-16982.71	ARIMA(2,1,1)	-13091.62
ARIMA(1,1,1)	-16970.99	ARIMA(3,1,2)	-13091.21
ARIMA(2,1,0)	-16982.36	ARIMA(2,1,3)	-13091.64
ARIMA(3,1,1)	-16996.22	ARIMA(1,1,1)	-13088.06
ARIMA(3,1,0)	-16984.81	ARIMA(1,1,3)	-13093.66
ARIMA(4,1,1)	-16993.27	ARIMA(3,1,1)	-13093.02
ARIMA(3,1,2)	-16996.82	ARIMA(3,1,3)	Inf
ARIMA(4,1,2)	-16991.65	ARIMA(2,1,2)	-13095.69
ARIMA(3,1,3)	-17001.98	ARIMA(1,1,2)	-13095.15
ARIMA(2,1,3)	-16992.11	ARIMA(2,1,1)	-13093.64
ARIMA(4,1,3)	-16990.5	ARIMA(3,1,2)	-13093.2
ARIMA(3,1,4)	-16998.78	ARIMA(2,1,3)	-13093.66
ARIMA(2,1,4)	-16992.88	ARIMA(1,1,1)	Inf
ARIMA(4,1,4)	-16994.43	ARIMA(1,1,3)	-13095.65
ARIMA(3,1,3)	-17003.8	ARIMA(3,1,1)	-13095.05
ARIMA(2,1,3)	-16994.1	ARIMA(3,1,3)	-13091.47
ARIMA(3,1,2)	-16998.65		
ARIMA(4,1,3)	-16992.29		
ARIMA(3,1,4)	-17000.71		
ARIMA(2,1,2)	-16982.88		
ARIMA(2,1,4)	-16994.75		
ARIMA(4,1,2)	-16993.41		
ARIMA(4,1,4)	-16996.14		

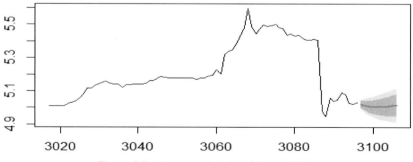

Forecasts from ARIMA(3,1,3)

Figure 9.9   Forecast plot of coal from ANNs.

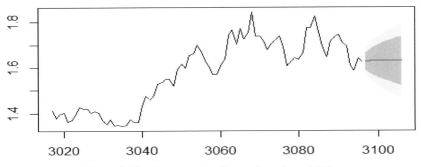

Figure 9.10 Forecast plot of natural gas from ANNs.

Table 9.6 Results of ANN-predictive model of coal.

Series: Coal
Model: NNAR(31,16)
Call: nnetar(*y* = coal)
Average of 20 networks, each of which is a 31-16-1 network with 529 weights

Table 9.7 Results of ANN-predictive model of natural gas

Series: NG
Model: NNAR(10,6)
Call: nnetar(*y* = NG)
Average of 20 networks, each of which is a 10-6-1 network with 73 weights

9.3.5 Discussion on the results of ANN predictive models

The results of ANNs (see Table 9.6) for coal show that the best-suited model is (31,16) and that for natural gas as seen in Table 9.7 is (10,6). The average number of networks in both commodities is 20 networks. The forecasting of 50-day prices of both commodities has been shown in Table 9.5. The plot of the forecasted prices of coal and natural gas has been shown in Figures 9.11 and 9.12. Finally, various errors such as mean error (ME), root mean squared error (RMSE), mean absolute error (MAE), mean percentage error (MPE), and mean absolute percentage error (MAPE) have been estimated. With these errors, we can check the robustness or accuracy of the model. Lower the error values, higher the accuracy of the model. According to the results of ME, RMSE, MAE, MPE, and MAPE, ANNs are the most robust model for both commodities between autoregressive and neural networks.

Forecasts from NNAR(31,16)

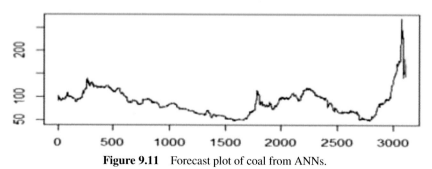

Figure 9.11　Forecast plot of coal from ANNs.

Forecasts from NNAR(10,6)

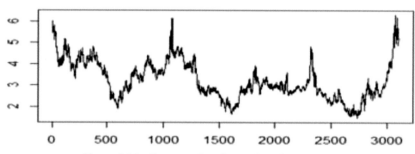

Figure 9.12　Forecast plot of natural gas from ANNs.

Table 9.8　Result of robustness of the model.

	ME	RMSE	MAE	MPE	MAPE
ARIMA_Coal	0.00013	0.0154	0.0068	0.00241	0.154
ARIMA_NG	−0.00003	0.0291	0.021	−0.0535	2.0253
ANNs_Coal	0.00000	0.0121	0.0023	0.0011	0.112
ANNs_NG	−0.0023	0.0112	0.011	−0.0002	1.012

9.4 Conclusion

For economies in general and developing economies in specific, energy is necessary for growth. Energy resources across the globe are gradually depleting due to easy exploitation by mining companies. The use of coal by the industries in an economy increases with the increase in crude oil prices. Developing countries are likely to observe explosive growth in renewable energy generation. India is moving forward in improving energy security through significant investments in strengthening its renewable capacities and allowing 100% FDI in the power sector. The Indian government aims to enhance natural gas share in the energy mix. India is progressing well toward achieving targets set by the United Nations for sustainable energy, and India is actively involved internationally in reducing its per capita carbon emissions. Increasing energy prices result in cost-push inflation in an economy. Thus, more accurate forecasting models are helpful. Studying and forecasting energy and raw material price movements is imperative and critical to get a timely indication for appropriate economic decisions. Thus, volatility estimation associated with energy prices is essential for improved regulators' decision-making accuracy.

With this background, this study contributes to the extant literature by estimating the prices for two important commodities – coal and natural gas – by deploying ARIMA and artificial neural network (ANN) techniques. From the preliminary analysis, we draw certain insights from investors' perspective, that the maximum return is given by natural gas (20%) followed by coal (14%). However, the standard deviation is high in natural gas (0.03), followed by coal (0.02). The natural gas returns are positively skewed, whereas returns of coal gas are negatively skewed. It means that natural gas has a high probability of getting positive returns, and coal has a probability of getting negative returns. Both series are integrated at I (1). We found that the best suitable ARIMA model with p,d,q specifications for NG and coal are 2,1,2 and 3,1,3, respectively.

Further, the ANN model has been applied in both variables. We found that the best predictable ANN model for coal is NNAR(31,16) and that for NG is NNAR(10,6). Finally, the study addresses the best suitable model between ARIMA and ANNs based on various errors, i.e., ME, RMSE, MPE, and MAPE. The lower the error values, the higher the accuracy of the model. The findings indicate that ANN is the most robust model to forecast the prices of these two commodities. The results can be helpful for investors looking to park money in these commodities. Further, the study can help policymakers formulate macroeconomic policies and management of inflation.

References

[1] Mohammadi, H. (2011). Long-run relations and short-run dynamics among coal, natural gas and oil prices. Applied Economics, 43 (2), 129-137.

[2] Kaufmann, R. K., & Hines, E. (2018). The effects of combined-cycle generation and hydraulic fracturing on the price for coal, oil, and natural gas: Implications for carbon taxes. Energy Policy, 118 (7), 603-611.

[3] Bachmeier, L. J., & Griffin, J. M. (2006). Testing for market integration: crude oil, coal, and natural gas. The Energy Journal, 27 (2), 55-71.

[4] Fan, J. H., & Todorova, N. (2017). Dynamics of China's carbon prices in the pilot trading phase. Applied Energy, 208 (27), 1452-1467.

[5] Guan, Q., & An, H. (2017). The exploration on the trade preferences of cooperation partners in four energy commodities' international trade: Crude oil, coal, natural gas and photovoltaic. Applied Energy, 203 (22), 154-163.

[6] Sarwar, S., Chen, W., & Waheed, R. (2017). Electricity consumption, oil price and economic growth: Global perspective. Renewable and Sustainable Energy Reviews, 76 (11), 9-18.

[7] Annual Energy Outlook. (2021). U.S. Energy Information Administration. https://www.eia.gov/outlooks/aeo/

[8] Zhang, C., Pu, C., Cao, R., Jiang, T., & Huang, G. (2019). The stability and roof-support optimization of roadways passing through unfavorable geological bodies using advanced detection and monitoring methods, among others, in the Sanmenxia Bauxite Mine in China's Henan Province. Bulletin of Engineering Geology and the Environment, 78 (7), 5087-5099.

[9] Prawisudha, P., Namioka, T., & Yoshikawa, K. (2012). Coal alternative fuel production from municipal solid wastes employing hydrothermal treatment. Applied Energy, 90 (1), 298-304.

[10] India 2020, Energy Policy Review. (2020). International Energy Agency. https://iea.blob.core.windows.net/assets/2571ae38-c895-430e-8b62-bc19019c6807/India_2020_Energy_Policy_Review.pdf

[11] Hassan, A. S., & Meyer, D. F. (2020). Analysis of the Non-Linear Effect of Petrol Price Changes on Inflation in South Africa. International Journal of Social Sciences and Humanity Studies, 12 (1), 34-49.

[12] Sharma, A., Rishad, A., & Gupta, S. (2019). Measuring the impact of oil prices and exchange rate shocks on Inflation: Evidence from India. Eurasian Journal of Business and Economics, 12 (24), 45-64.

[13] Chen, Y. C., Rogoff, K. S., & Rossi, B. (2010). Can exchange rates forecast commodity prices?. The Quarterly Journal of Economics, 125 (3), 1145-1194.

[14] Sun, W., & De, K. (2019). Real exchange rate, monetary policy, and the US economy: Evidence from a FAVAR model. Economic Inquiry, 57 (1), 552-568.

[15] Alameer, Z., Abd Elaziz, M., Ewees, A. A., Ye, H., & Jianhua, Z. (2019). Forecasting gold price fluctuations using improved multilayer perceptron neural network and whale optimization algorithm. Resources Policy, 61 (2), 250-260.

[16] Alameer, Z., Abd Elaziz, M., Ewees, A. A., Ye, H., & Jianhua, Z. (2019)a. Forecasting copper prices using hybrid adaptive neuro-fuzzy inference system and genetic algorithms. Natural Resources Research, 28 (4), 1385-1401.

[17] Bildirici, M. E., & Sonustun, F. O. (2018). The effects of oil and gold prices on oil-exporting countries. Energy strategy reviews, 22 (4), 290-302.

[18] Jain, A., & Biswal, P. C. (2016). Dynamic linkages among oil price, gold price, exchange rate, and stock market in India. Resources Policy, 49 (3), 179-185.

[19] Sari, R., Hammoudeh, S., & Soytas, U. (2010). Dynamics of oil price, precious metal prices, and exchange rate. Energy Economics, 32 (2), 351-362.

[20] Singhal, S., Choudhary, S., & Biswal, P. C. (2019). Return and volatility linkages among International crude oil price, gold price, exchange rate and stock markets: Evidence from Mexico. Resources Policy, 60 (1), 255-261.

[21] Soytas, U., Sari, R., Hammoudeh, S., & Hacihasanoglu, E. (2009). World oil prices, precious metal prices and macroeconomy in Turkey. Energy Policy, 37 (12), 5557-5566.

[22] Haque, M. A., Topal, E., & Lilford, E. (2017). Evaluation of a mining project under the joint effect of commodity price and exchange rate uncertainties using real options valuation. The Engineering Economist, 62 (3), 231-253.

[23] Dong, B., Li, X., & Lin, B. (2010). Forecasting Long-Run Coal Price in China: A Shifting Trend Time-Series Approach. Review of Development Economics, 14(3), 499-519.

[24] Xue, G., & Sriboonchitta, S. (2014). Co-movement of prices of energy and agricultural commodities in biofuel Era: a period-GARCH Copula approach. Modeling Dependence in Econometrics, 251, 505-519.

[25] Lago, J., De Ridder, F., & De Schutter, B. (2018). Forecasting spot electricity prices: Deep learning approaches and empirical comparison of traditional algorithms. Applied Energy, 221(13), 386-405.

[26] Herrera, G. P., Constantino, M., Tabak, B. M., Pistori, H., Su, J. J., & Naranpanawa, A. (2019). Long-term forecast of energy commodities price using machine learning. Energy, 179(14), 214-221.

[27] Matyjaszek, M., Fernández, P. R., Krzemień, A., Wodarski, K., & Valverde, G. F. (2019). Forecasting coking coal prices by means of ARIMA models and neural networks, considering the transgenic time series theory. Resources Policy, 61(2), 283-292.

[28] Wang, B., & Wang, J. (2019). Energy futures prices forecasting by novel DPFWR neural network and DS-CID evaluation. Neurocomputing, 338(15), 1-15.

[29] Krzemień, A., Fernández, P. R., Sánchez, A. S., & Álvarez, I. D. (2016). Beyond the pan-european standard for reporting of exploration results, mineral resources and reserves. Resources Policy, 49(3), 81-91.

[30] Kriechbaumer, T., Angus, A., Parsons, D., & Casado, M. R. (2014). An improved wavelet–ARIMA approach for forecasting metal prices. Resources Policy, 39(1), 32-41.

[31] Dooley, G., & Lenihan, H. (2005). An assessment of time series methods in metal price forecasting. Resources Policy, 30(3), 208-217.

[32] Alameer, Z., Fathalla, A., Li, K., Ye, H., & Jianhua, Z. (2020). Multistep-ahead forecasting of coal prices using a hybrid deep learning model. Resources Policy, 65(1), 101588.

[33] Han, M., Ding, L., Zhao, X., & Kang, W. (2019). Forecasting carbon prices in the Shenzhen market, China: The role of mixed-frequency factors. Energy, 171(6), 69-76.

[34] Matyjaszek, M., Fidalgo Valverde, G., Krzemień, A., Wodarski, K., & Riesgo Fernández, P. (2020). Optimising predictor variables in artificial neural networks when forecasting raw material prices for energy production. Energies, 13(8), 2017.

[35] Yang, Y., Zheng, X., & Sun, Z. (2020). Coal resource security assessment in China: A study using entropy-weight-based TOPSIS and BP neural network. Sustainability, 12(6), 2294.

[36] Zhang, J., Li, D., Hao, Y., & Tan, Z. (2018). A hybrid model using signal processing technology, econometric models and neural network for carbon spot price forecasting. Journal of Cleaner Production, 204(35), 958-964.

[37] Bento, P. M. R., Pombo, J. A. N., Calado, M. R. A., & Mariano, S. J. P. S. (2018). A bat optimized neural network and wavelet transform approach for short-term price forecasting. Applied energy, 210(2), 88-97.

[38] Yu, L., Zhang, X., & Wang, S. (2017). Assessing potentiality of support vector machine method in crude oil price forecasting. EURASIA Journal of Mathematics, Science and Technology Education, 13(12), 7893-7904.

[39] Čeperić, E., Žiković, S., & Čeperić, V. (2017). Short-term forecasting of natural gas prices using machine learning and feature selection algorithms. Energy, 140(23), 893-900.

[40] Cuaresma, J. C., Hlouskova, J., Kossmeier, S., & Obersteiner, M. (2004). Forecasting electricity spot-prices using linear univariate time-series models. Applied Energy, 77(1), 87-106.

[41] Kim, S., Ko, W., Nam, H., Kim, C., Chung, Y., & Bang, S. (2017). Statistical model for forecasting uranium prices to estimate the nuclear fuel cycle cost. Nuclear Engineering and Technology, 49(5), 1063-1070.

[42] Matyjaszek, M., Wodarski, K., Krzemień, A., García-Miranda, C. E., & Sánchez, A. S. (2018). Coking coal mining investment: Boosting European Union's raw materials initiative. Resources Policy, 57(2), 88-97.

[43] He, K., Chen, Y., & Tso, G. K. (2017). Price forecasting in the precious metal market: A multivariate EMD denoising approach. Resources Policy, 54(4), 9-24.

[44] Lasheras, F. S., de Cos Juez, F. J., Sánchez, A. S., Krzemień, A., & Fernández, P. R. (2015). Forecasting the COMEX copper spot price by means of neural networks and ARIMA models. Resources Policy, 45(3), 37-43.

[45] Guo, X., Shi, J., & Ren, D. (2016). Coal price forecasting and structural analysis in China. Discrete Dynamics in Nature and Society, 2016, 1-7.

[46] Ming, Z., Shulei, L., Song, X., Xiaoli, Z., & Lingyun, L. (2016). Prediction of China's coal price during Twelfth Five Year Plan period. Energy Sources, Part B: Economics, Planning, and Policy, 11(6), 511-517.

[47] Kong, Z., Dong, X., & Jiang, Q. (2019). Forecasting the development of China's coal-to-liquid industry under security, economic and environmental constraints. Energy Economics, 80(4), 253-266.

[48] Mengdi, Z., & Yong, S. K. (2018). Forecasting the Carbon Price in China Pilot Emission Trading Scheme: A Structural Time Series Approach. In The State of China's State Capitalism, 117-139.

[49] Aggarwal, V., Doifode, A., & Tiwary, M. K. (2021). Do Lower Foreign Flows and Higher Domestic Flows Reduce Indian Equity Market Volatility? Vision. https://doi.org/10.1177/0972262921990981

[50] Rastogi, S., Doifode, A., Kanoujiya, J., & Singh, S. P. (2021). Volatility integration of gold and crude oil prices with the interest rates in India. South Asian Journal of Business Studies. (In Press)

[51] Aggarwal, V. (2021). Optimum investor portfolio allocation in new age digital assets. International Journal of Innovation Science. (In Press)

[52] Aggarwal, V., & Tiwary, M. K. (2022). Impact of State Intervention in Oil Pricing on Stock Market Volatility. Economic and political weekly, 57(6), 51-58.

[53] Chu, J., Liu, X., Zhang, Z., Zhang, Y., & He, M. (2021). A novel method overcomeing overfitting of artificial neural network for accurate prediction: Application on thermophysical property of natural gas. Case Studies in Thermal Engineering, 28, 101406.

[54] Al-AbdulJabbar, A., Mahmoud, A. A., & Elkatatny, S. (2021). Artificial neural network model for real-time prediction of the rate of penetration while horizontally drilling natural gas-bearing sandstone formations. Arabian Journal of Geosciences, 14(2), 1-14.

[55] Liu, J., Huang, Q., Ulishney, C., & Dumitrescu, C. E. (2022). Comparison of Random Forest and Neural Network in Modeling the Performance and Emissions of a Natural Gas Spark Ignition Engine. Journal of Energy Resources Technology, 144(3).

Index

About the Editors

Dr. Neelu Nagpal (Senior Member, IEEE), an Associate Professor at GGSIP University, has been working for the last 16 years in the EEE Department of Maharaja Agrasen Institute of Technology, Delhi, India. Her Ph.D. in Electrical Engineering was accomplished at Delhi Technological University, Delhi. She was the recipient of commendable research award from Delhi Technological University during her Ph.D. course. She completed her Masters with distinction from Delhi University in Control and Instrumentation specialization. Besides having 22 years of experience in teaching, she has 5 years of industrial experience and a lot more educational contributions to her name. She has 20 research publications in top-tier journals and conferences and has a grant of one Australian patent. Her area of research is investigations into the dynamics, control, and estimation in the field of nonlinear stochastic systems (power systems and robotics), smart grid technologies, and artificial intelligence. She has been involved in many reputed conferences in various capacities such as advisor, reviewer, session-chair, track-chair, organizing committee member, and publication chair (IEEE conference-ICIERA-2021). She is an IEEE Smart Cities Ambassador, 2022.

Professor Hassan Haes Alhelou(Senior Member, IEEE) is with Monash University, Australia. He is also a Professor at Tishreen University in Syria, and a consultant with Sultan Qaboos University in Oman. Previously, He was with UCD-Ireland and IUT-Iran. He was included in the 2018 and 2019 Publons list of the top 1% best reviewers in the world. He is the recipient of the Outstanding Reviewer Award from many journals, e.g., ECM, ISA Transactions, and Applied Energy; and the recipient of the best young researcher in the Arab Student Forum Creative among 61 researchers from 16 countries at Alexandria University, Egypt, 2011. He also received the Excellent Paper Award 2021 from IEEE CSEE Journal of Power and Energy Systems. He has published more than 200 research papers in high-quality peer-reviewed journals. His research papers have received 2850 citations with an H-index of 26 and an i-index of 56. He has authored/edited 15 books published by reputed publishers. He serves as an editor for a number of prestigious journals. He has participated in more than 15 international industrial projects over the globe. His major research interests are power system security and dynamics, power system cybersecurity, power system operation, and control, dynamic state estimation, and smart grids.

Professor Pierluigi Siano (M'09-SM'14) received the M.Sc. degree in electronic engineering and the Ph.D. degree in information and electrical engineering from the University of Salerno, Salerno, Italy, in 2001 and 2006, respectively. He is a Professor and Scientific Director of the Smart Grids and Smart Cities Laboratory with the Department of Management & Innovation Systems, University of Salerno. Since 2021 he has been a Distinguished Visiting Professor in the Department of Electrical & Electronic Engineering Science, University of Johannesburg and since 2023 he has been Fellow of the Distinguished Scientist Fellowship Program with the Department of Electrical Engineering, College of Engineering, King Saud University. His research activities are centered on demand response, energy management, the integration of distributed energy resources in smart grids, electricity markets, and planning and management of power systems. In these research fields, he has co-authored more than 700 articles including more than 410 international journals that received in Scopus more than 17000 citations with an H-index equal to 63. In the period 2019-2022 he has been awarded as a Highly Cited Researcher in Engineering by Web of Science Group. He has been the Chair of the IES TC on Smart Grids. He is Editor for the Power & Energy Society Section of IEEE Access, IEEE TRANSACTIONS ON POWER SYSTEMS, IEEE TRANSACTIONS ON INDUSTRIAL INFORMATICS, IEEE TRANSACTIONS ON INDUSTRIAL ELECTRONICS, IEEE SYSTEMS.

Professor Sanjeevikumar Padmanaban (Member'12–Senior Member'15-IEEE) received his bachelor's, master and Ph.D. degrees in electrical engineering from the University of Madras, India, 2002, Pondicherry University, India, 2006, and the University of Bologna, Italy, 2012. Since 2018, he has been with the Department of Energy Technology, Aalborg University, Esbjerg, Denmark, as an Assistant Professor. He has authored 300 plus scientific papers and has been involved in various capacities in many international conferences, including for IEEE and IET. He has received the best paper award from IET-SEISCON'13, IET-CEAT'16, and ETAEERE'16 sponsored Springer book publication series (five papers). He serves as an

Editor/Associate Editor/Editorial Board of many refereed journals in particular the IEEE Systems Journal, IEEE Access, the IET Power Electronics, the subject editor of IET Renewable Power Generation, the subject editor of IET Generation, Transmission, and Distribution, Journal of Power Electronics (Korea) and FACETS (Canada). He is a fellow of the Institution of Engineers – FIE '18 (India) and a fellow of the Institution of Telecommunication and Electronics Engineers – FIETE'18 (India).

Dr. D. Lakshmi is presently designated as a Senior Associate Professor in the School of Computing Science and Engineering (SCSE) and Assistant Director, Centre for Innovation in Teaching & Learning (CITL) at VIT Bhopal. She has 25 years of teaching experience. She has given innumerable guest lectures, acted as a session chair, and has been invited as a keynote speaker. She has conducted FDPs that cover approximately 50,000 plus faculty members including JNTU, TEQIP, SERB, SWAYAM, DST, AICTE, MHRD, ATAL, ISTE sponsored, and self-financed workshops across India. She has given 17 international conference presentations, has written 17 international journal papers including for SCOPUS, WOS, SCIE, SCI, 2 book chapters, 1 German patent, 3 Indian patents granted, 8 are yet to be granted, 2 Indian Copyrights granted, 4 Australian patents are granted, 8 more Indian patents are published and waiting for grant. A total of 18 patents are in various states. She has won two Best Paper awards at international conferences, and published a book on "Theory of Computation" in the year 2003 and "Leading Education in the Age of Disruption" in the year 2021.